MEMOIRS
of the
American Mathematical Society

Number 447

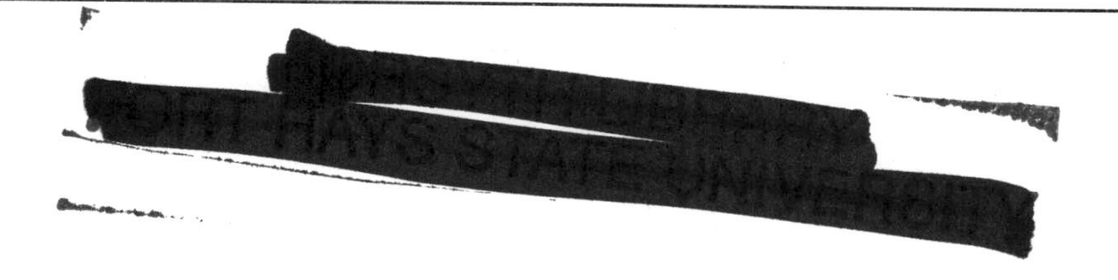

No Nine Neighborly Tetrahedra Exist

Joseph Zaks

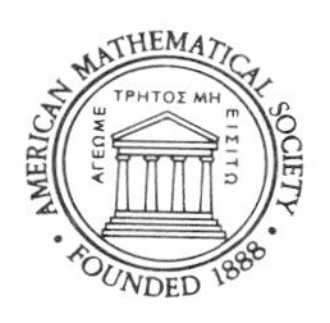

May 1991 • Volume 91 • Number 447 (end of volume) • ISSN 0065-9266

American Mathematical Society
Providence, Rhode Island

1980 *Mathematics Subject Classification* (1985 *Revision*).
Primary 52A25, 52A15, 51-04, 51M20; Secondary 05C50.

Library of Congress Cataloging-in-Publication Data

Zaks, Joseph, 1940–
No nine neighborly tetrahedra exist/Joseph Zaks.
p. cm. – (Memoirs of the American Mathematical Society, ISSN 0065-9266; no. 447)
"May 1991, volume 91, number 447 (end of volume)."
Includes bibliographical references and index.
ISBN 0-8218-2517-8
1. Tetrahedra. I. Title. II. Series.
QA561.Z34 1991 91-11255
516.3′3–dc20 CIP

Subscriptions and orders for publications of the American Mathematical Society should be addressed to American Mathematical Society, Box 1571, Annex Station, Providence, RI 02901-1571. *All orders must be accompanied by payment.* Other correspondence should be addressed to Box 6248, Providence, RI 02940-6248.

SUBSCRIPTION INFORMATION. The 1991 subscription begins with Number 438 and consists of six mailings, each containing one or more numbers. Subscription prices for 1991 are $270 list, $216 institutional member. A late charge of 10% of the subscription price will be imposed on orders received from nonmembers after January 1 of the subscription year. Subscribers outside the United States and India must pay a postage surcharge of $25; subscribers in India must pay a postage surcharge of $43. Expedited delivery to destinations in North America $30; elsewhere $82. Each number may be ordered separately; *please specify number* when ordering an individual number. For prices and titles of recently released numbers, see the New Publications sections of the NOTICES of the American Mathematical Society.

BACK NUMBER INFORMATION. For back issues see the AMS Catalogue of Publications.

MEMOIRS of the American Mathematical Society (ISSN 0065-9266) is published bimonthly (each volume consisting usually of more than one number) by the American Mathematical Society at 201 Charles Street, Providence, Rhode Island 02904-2213. Second Class postage paid at Providence, Rhode Island 02940-6248. Postmaster: Send address changes to Memoirs of the American Mathematical Society, American Mathematical Society, Box 6248, Providence, RI 02940-6248.

Printed in the United States of America.
The paper used in this book is acid-free and falls within the guidelines
established to ensure permanence and durability. ♾

10 9 8 7 6 5 4 3 2 1 95 94 93 92 91

Table of content

Abstract

A long standing conjecture of Bagemihl (1956) states that there can be at most eight tetrahedra in 3-space, such that every two of them meet in a two dimensional set. We settle this conjecture affirmatively.

We get some information on families of similar nature, consisting of eight tetrahedra. We present a joint result, showing that there can be at most fourteen tetrahedra in 3-space, such that for every two of them there is a plane which separates them and contains a facet of each one of them.

Key words and phrases: Tetrahedron, neighborly family, nearly-neighborly family, complete bipartite graph.

Dedicated to the memory of my parents

Dvora (1913-1987) and **Izhak Meir** (1913-1990) **Zaks**

Acknowledgement

We wish to thank all those who partially supported our research on Bagemihl's conjecture during the years we have been working on it. Most of the results were obtained while the author was visiting A. Kotzig and G. Sabidussi at the University of Montreal, A. Bondy and other colleagues at the C. & O. Department, University of Waterloo and R.K. Guy at the University of Calgary; these visits were supported by the hosts' NSERC grants.

We wish to thank P. Laufer for some computations at an early stage of the research, and to thank N. Linial for some helpful conversations concerning the Graham-Pollak Theorem.

Special thanks go to my son Ayal Zaks for the many fruitful discussions on the computational aspects of our searches, and for his help in getting the many graphical representations needed during the work.

We highly appreciate the effort of the referees and their many valuable comments and suggestions.

Introduction

A family of convex d-polytopes in the d-space is called *neighborly* [1,5,8,9,10,12,14,17-22] if every pair of its members intersect in a (d-1)-dimensional set; such an intersection lies in a hyperplane which separates the pair and contains a facet of each one of them.

In studying possible extensions of the Four-Color Conjecture to E^3, Tietze [15] in 1905 and Besicovitch [3] in 1947 gave an example of an infinite neighborly family of convex 3-polytopes in E^3. In 1956, Bagemihl [1] restricted the attention to neighborly families of tetrahedra. Bagemihl gave the example of eight neighborly tetrahedra, shown here in Figure 1. All the tetrahedra have a facet on a common plane, which separates four of them from the remaining four. Each one of the two quadruples shares a common vertex in the open half-space determined by the said plane.

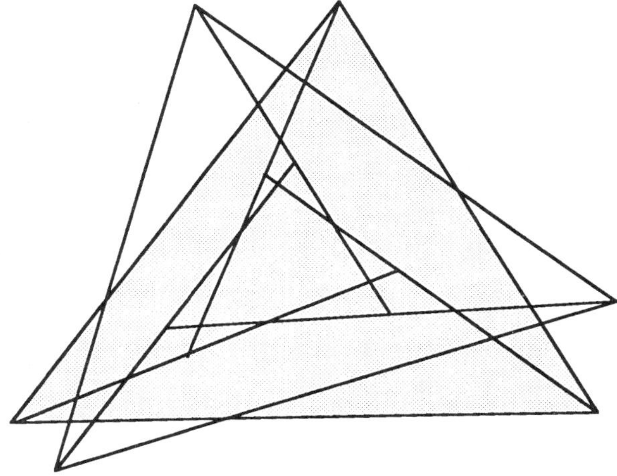

Figure 1. The bases of eight neighborly tetrahedra. Four of the tetrahedra are above the plane, and the remaining four are below it.

Another example of eight neighborly tetrahedra was given by S. Wilson and the author [17]; here, too, there are two quadruples of tetrahedra separated by a plane which contains a facet of each one of the tetrahedra; this example is shown in Figure 2.

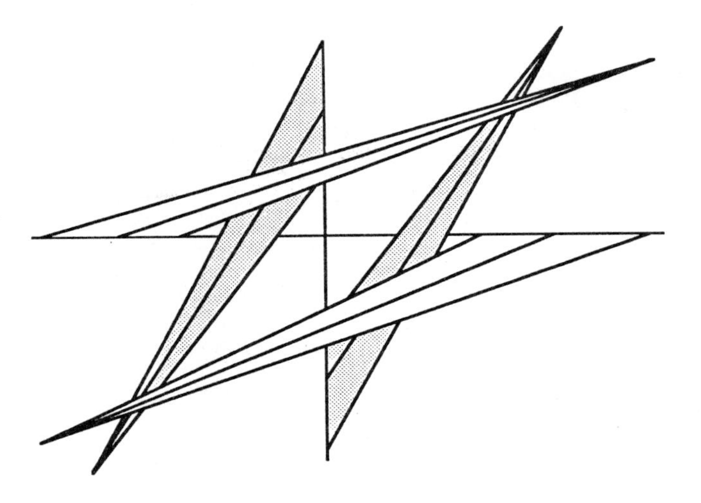

Figure 2. The bases of eight tetrahedra, in another example of a neighborly family, taken from [17].

Baston stated an argument which implies that a neighborly family of tetrahedra contains at most seventeen members. He conjectured that the maximum number of tetrahedra in a neighborly family is eight. Baston [2] showed in 1965 that this maximum is at most nine, and he repeated Bagemihl's conjecture that the maximum is eight.

Bagemihl's conjecture had been repeatedly mentioned in the literature: by Danzer, Grunbaum and Klee [5] in 1963, by Grunbaum [8] in 1967 and in particular by Klee [10] in 1969 (it is also mentioned by Klee in his research-film called "Shapes of the Future", part two, produced by the M.A.A. at about 1972); it is also mentioned by Perles [12], by kassem [9] and by us [17,18].

In this article we present a solution to Bagemihl's conjecture, by proving the following.

Main Theorem: *There exist no neighborly families in E^3, consisting of nine tetrahedra.*

This result has been announced in [21]. For other results, dealing with neighborly families in E^d, $d \geq 3$, see [19, 20, 22]; for related results, see [14, 23, 6].

Our proof depends, in a very strong sense, on Baston's [2] result and technique. We complement it with a few new ideas from Combinatorial geometry, graph theory and a number of exhaustive searches by computer, done on our private Commodore 64.

The proof of the Main Theorem is presented in the first nine chapters. In chapter 1, assuming that there exists a neighborly family F consisting of nine tetrahedra, we introduce its Baston matrix B(F), the corresponding decomposition of K_9 into complete bipartite graphs, the Diophantine system of equations and its twenty four solutions. In Chapter 2 we give direct reasons showing that *ten* of the solutions lead to contradictions. In Chapter 3 we introduce the notion of a *type* of a tetrahedron in F, to be used later. In Chapter 4 we present a well-known lemma from Linear Programming, which will be repeatedly used in showing that certain $\{0, \pm 1\}$-matrices cannot be the Baston matrix B(f) of F. This leads to a few properties of the possible matrices, which are translated into properties of the decomposition of K_9. These properties will ease the task of search by a computer. Chapters 5 - 9 treat separate cases of the computer search, where all the corresponding decompositions of K_9 are produced (up to combinatorial type), and each one of them is shown to lead to a contradiction. In Chapter 10 we treat the problem of the maximum number of combinatorially different neighborly families of eight tetrahedra in E^3, and we close in Chapter 11 by reproducing the proof, from [6], that there can be at most *fourteen* nearly-neighborly tetrahedra in E^3.

Chapter 1

Our proof of the Main Theorem starts by assuming, on the contrary, that there exists a neighborly family F in E^3, consisting of nine tetrahedra $P_1, P_2, \dots, P_9$. Let $\{H_1, \dots, H_t\}$ denote the collection of all the planes in E^3 which contain facets of members of F. Let H^+_j and H^-_j denote the two closed half-spaces, determined by H_j, for all j.

Following Baston [2] (see also [17, 12, 20, 21, 23]), let $B(F) = (b_{i,j})$ be the Baston matrix of F, defined by

$$b_{i,j} = \begin{cases} +1 & \text{if } H_j \text{ contains a facet of } P_i \text{ and } P_i \subset H_j^+, \\ -1 & \text{if } H_j \text{ contains a facet of } P_i \text{ and } P_i \subset H_j^-, \\ 0 & \text{otherwise,} \end{cases}$$

for all $1 \le i \le 9$ and $1 \le j \le t$.

The following properties hold:

(1) Each row of B(F) has precisely four nonzero terms.

(2) For every pair of row-indices i and j, there exists a unique column-index k, such that $b_{ik} \cdot b_{jk} = -1$.

Property (1) follows from the fact that every tetrahedron has four facets, while property (2) is implied by the neighborliness of F, see [2,17].

Following [17, 20, 21], let $x_{i,j}$, $i \leq j$, denote the number of columns of B(F), consisting of i nonzero terms of one sign and j nonzero terms of the opposite sign, besides zeros. If $x_{0,j} > 0$, for $j > 1$, we cut j-1 of the tetrahedra, having a facet on a common plane, by parallel planes, such that the neighborliness is preserved and such that the j facets lie on different planes. We may assume, therefore, that $x_{0,j} > 0$ implies that $j = 1$. The quantity $x_{0,1}$ is just the total number of *free* facets (for definition, see [17]) of members of F.

The following are known:

(3) $x_{i,j} > 0$ implies $j \leq 3$.

(4) $x_{3,3} = 0$.

Property (3) is Lemma 16 in [2, p.56], while property (4) is one of the main steps in [2], proved as Theorem 7 in p.157, there.

A member of F is said to be of *type* (p, q, r, s), $p \geq q \geq r \geq s$, if it touches (in a two-dimensional set) p other members of F in one facet, q of them in another facets, etc. Thus, there are just four possible types: (3, 3, 2, 0), (3, 3, 1, 1), (3, 2, 2, 1) and (2, 2, 2, 2). Let α, β, γ and δ denote the number of members of F of types (3, 3, 2, 0), (3, 3, 1, 1), (3, 2, 2, 1) and (2, 2, 2, 2), respectively. Using Baston's nomenclature, α is the number of "naiks", β - of "dhoats", γ - of "thaiks" and δ - of "chardhos".

Clearly, only the members of F of type (3, 3, 2, 0) have free facets, one each.

Therefore

(5) $\alpha = x_{0,1}$.

Theorem 9 of [2, p.186] implies

(6) $0 \leq x_{0,1} \leq 2$.

By counting the total number of the nonzero terms of B(F), in two ways (rows and columns), we easily get (see [17]) the following.

(7) $$\sum_{i,j} (i+j)x_{i,j} = 36 .$$

Property (2) of B(F) implies that every two rows contributes exactly one minor of the form

$\begin{pmatrix} 1 \\ -1 \end{pmatrix}$ or $\begin{pmatrix} -1 \\ 1 \end{pmatrix}$,

thus there are 36 (= 9·8/2) such minors. On the other hand, each column counted by $x_{i,j}$ contributes $i \cdot j$ such minors, hence we have

(8) $$\sum_{i,j \geq 1} ijx_{i,j} = 36 .$$

It is worth mentioning, as we already did in [17], that the system of equations (7) and (8) is equivalent to the relations obtained by Baston [2] in connection with his notion of "surplus".

Let us define a (possibly multi-) graph G, having the nine vertices $\{1, 2, ..., 9\}$. A vertex i is connected by an edge to a vertex j whenever there exists a k, $1 \le k \le t$, such that $b_{ik} \cdot b_{jk} = -1$. By (2), it follows that G is the complete graph K_9. Considering the edges which are contributed by any one column of $B(F)$, it follows that K_9 has a decomposition into the sum, over all i and j, of $x_{i,j}$ copies of the complete bipartite graph $K_{i,j}$ having i vertices of one class and j vertices of the other class. Thus, we have the following edge-disjoint decomposition:

$$K_9 = \sum_{1 \le i \le j \le 3} x_{i,j} K_{i,j} . \tag{9}$$

The Graham-Pollak Theorem [7] (for simpler proofs, see [11, 16]; for extensions, see [13, 23, 6]) implies that

$$\sum_{1 \le i \le j \le 3} x_{i,j} \ge 8 . \tag{10}$$

The Diophantine system (3), (4), (6), (7) and (8) has 24 possible solutions, given in Table 1. The idea of the proof is to show that each one of these solutions, which will be called here "cases", leads to contradictions.

Case #	$x_{0,1}$	$x_{1,1}$	$x_{1,2}$	$x_{1,3}$	$x_{2,2}$	$x_{2,3}$	Impossible by
1	0	0	0	0	9	0	Baston [2], Theorem 10, p.196.
2	0	0	1	0	7	1	Chapter 5.
3	0	0	2	0	5	2	Chapter 6.
4	0	1	0	1	5	2	Theorem 3.
5	2	0	0	0	6	2	Theorem 1.
6	0	0	3	0	3	3	Chapter 7.
7	0	1	1	1	3	3	Theorem 3.
8	1	0	0	2	3	3	Baston's Theorem 8; Theorem 2.
9	1	2	0	0	4	3	Chapter 7.
10	2	0	1	0	4	3	"
11	0	0	0	4	0	4	Theorem 2.
12	0	0	4	0	1	4	Chapter 8.
13	0	1	2	1	1	4	Theorem 3.
14	0	2	0	2	1	4	Theorem 3.
15	1	0	1	2	1	4	Theorem 2.
16	0	4	0	0	2	4	Chapter 8.
17	1	2	1	0	2	4	"
18	2	0	2	0	2	4	"
19	2	1	0	1	2	4	Theorem 4.
20	0	4	1	0	0	5	Chapter 9.
21	1	2	2	0	0	5	"
22	2	0	3	0	0	5	"
23	1	3	0	1	0	5	"
24	2	1	1	1	0	5	Theorem 4.

Table 1. The 24 solution of the equations (3, 4, 6, 7, 8).

Baston proved that case #1 (of our table 1) is impossible (Theorem 10, p. 196 of [2]). He observed that all the facets of members of F cannot lie on just eight planes (Theorem 8, p.186 of [2]), thus case #11 is also impossible. We shall give a simpler proof that case #11 is impossible .

Chapter 2

In this chapter we prove that the ten cases #4, 5, 7, 8, 11, 13, 14, 15, 19 and #24 of Table 1 lead to contradictions. We split the proofs, according to the reasons used, as follows.

Theorem 1: *Case #5 is impossible.*

Theorem 2: *Cases #8 , 11 and #15 are impossible.*

Theorem 3: *Cases #4 , 7 , 13 and #14 are impossible.*

Theorem 4: *Cases #19 and #24 are impossible.*

Proof of Theorem 1: The corresponding decomposition (9) of K_9 in this case is $K_9 = 2K_{2,3} + 6K_{2,2}$. The only vertices of odd degrees in K_9, $K_{2,3}$ and $K_{2,2}$ are the two vertices of $K_{2,3}$ which are 3-valent. Thus, the two couples of 3-valent vertices in the two $K_{2,3}$'s must coincide, so that the two $K_{2,3}$ actually form a $K_{2,6}$. It follows that K_9 has a decomposition of the form $K_{2,6} + 6K_{2,2}$, contrary to the Graham-Pollak Theorem [7] (see also property (10)). Thus the decomposition $K_9 = 2K_{2,3} + 6K_{2,2}$ is impossible, therefore case #5 is impossible. This completes the proof of Theorem 1.

For the proofs of Theorems 2 - 4 we need the following.

Lemma 1: The maximum possible number of *bounded* parts, determined by n planes in E^3, is $(n-1)(n-2)(n-3)/6$.

The *total* maximum possible number of parts, bounded as well as unbounded, determined by n hyperplanes in E^d is well known; see, for example, [4].

Proof of Lemma 1: Let $b(d, n)$ denote the maximum possible number of bounded parts, determined by n hyperplanes in E^d. We may assume that the hyperplanes are such that no two of them are parallel. It follows easily, by deleting any one hyperplane, that $b(d, n) = b(d-1, n-1) + b(d, n-1)$; therefore $b(d, n)$ is a polynomial in n of degree d. It takes at least d+1 hyperplane to determine one bounded part (a d-simplex), therefore $b(d, n) = 0$ for all n, $1 \leq n \leq d$, and $b(d, d+1) = 1$. Hence $b(d, n) = (n-1)(n-2)...(n-d)/d!$.

Lemma 1 treats the particular case in which $d = 3$.

Lemma 2: If column j of B(F) has nonzero terms of opposite signs, and if $b_{ij} = 0$, then the plane H_j passes through the interior of the tetrahedron P_i, i.e., H_j cuts P_i.

Proof of Lemma 2: Suppose $b_{rj} = 1$, $b_{sj} = -1$ and $b_{ij} = 0$. Therefore the tetrahedron P_i has no facets on the plane H_j, yet it meets P_r and P_s in 2-dimensional sets which are not included in H_j. Therefore $H^+_j \backslash H_j \supset \text{relint}(P_i \cap P_r)$ and $H^-_j \backslash H_j \supset \text{relint}(P_i \cap P_s)$, where *relint*(x) denotes the relative interior of x, see [8]. Thus, both $P_i \cap (H^+_j \backslash H_j) \neq \emptyset$ and $P_i \cap (H^-_j \backslash H_j) \neq \emptyset$, as claimed.

We proceed to the

Proof of Theorem 2: In case #11, $x_{0,1} = 0$, therefore by Lemma 2 every plane among H_1, ..., H_8 cuts all the tetrahedra which do not have a facet on that plane. Therefore every tetrahedron is cut by four (= 8 - 4) of the planes, giving rise to at least five bounded parts in every tetrahedron. The nine tetrahedra are cut by the eight planes into at least 45 (= 9·5) bounded parts. However, according to Lemma 1, eight planes can determine at most 35 bounded parts, therefore case #11 is impossible.

In cases #8 and #15, $x_{0,1} = 1$ and there are altogether nine planes, one of which contains just one free facet of a tetrahedron P_o. Eight of the tetrahedra, except for P_o, are cut by exactly four of the nine planes (the nine planes minus the four which contain facets of a tetrahedron and also minus the one plane which contains the free facet of P_o). These eight tetrahedra are cut into at least five bounded parts each, yielding at least 40 ($= 8 \cdot 5$) bounded parts. Let P_o be cut into k bounded parts by the nine planes. We delete P_o together with the plane which contains the free facet of P_o. We lose at most k bounded parts which were inside P_o (they may be enlarged into unbounded parts). As a result, we get eight planes which determine at least forty bounded parts, contrary to the maximum of thirty five parts, established in Lemma 1.

This completes the proof of Theorem 2.

For the proof of Theorems 3 and 4 we need the following.

Lemma 3: If T_1 and T_2 are two neighborly tetrahedra in E^3, where

$$T_1 = H^+ \cap \bigcap_{i=1}^{3} H_i^+ \quad \text{and} \quad T_2 = H^- \cap \bigcap_{i=4}^{6} H_i^+$$

represent them as the intersection of four closed half-spaces, then the set denoted by $T_1 * T_2$ (see Figure 3) and defined by

$$T_1 * T_2 = \bigcap_{i=1}^{6} H_i^+ ,$$

is a convex 3-polytope.

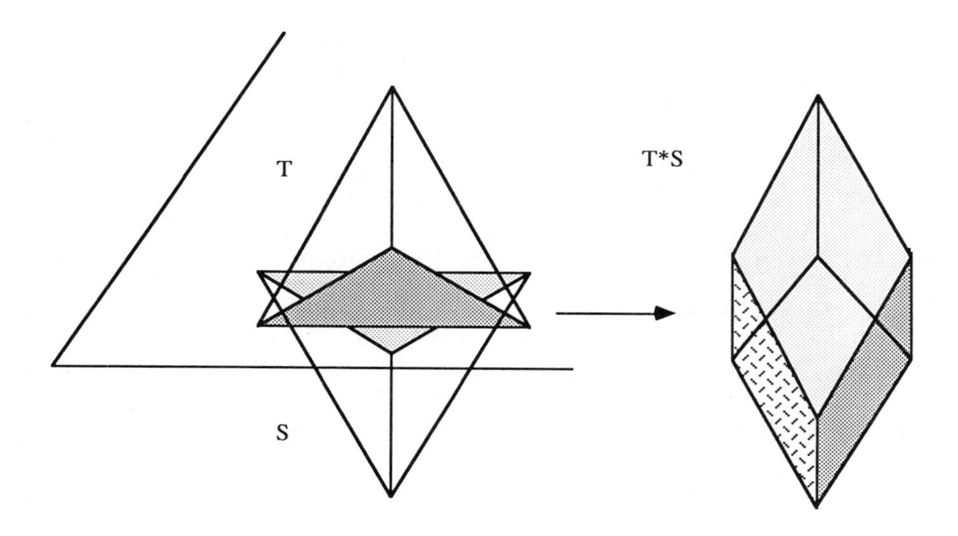

Figure 3. Two neighborly tetrahedra T and S, yielding the convex polytope T*S.

Proof of Lemma 3: $T_1 \cap T_2$ is 2-dimensional, by the neighborliness. For any point z in relint($T_1 \cap T_2$), there exists a small ball centered at z and inside $T_1 \cup T_2$. This ball is in $T_1 * T_2$, since it is in H^+_i for all i, $1 \leq i \leq 6$. Therefore $T_1 * T_2$ is 3-dimensional. It is clearly polyhedral and convex. The boundedness of $T_1 * T_2$ follows from

$$T_1 * T_2 = \bigcap_{i=1}^{6} H_i^+ = (H^+ \cup H^-) \cap (\bigcap_{i=1}^{6} H_i^+) = [H^+ \cap (\bigcap_{i=1}^{6} H_i^+)] \cup [H^- \cap (\bigcap_{i=1}^{6} H_i^+)] \subseteq T_1 \cup T_2 .$$

Therefore $T_1 * T_2$ is a convex 3-polytope, as claimed.

Proof of Theorem 3: In cases #4, 7, 13 and #14, $x_{0,1} = 0$, $x_{1,1} \geq 1$ and there are altogether nine planes. Let P_i and P_j be the two tetrahedra which meet in one of the planes, counted by $x_{1,1}$. We delete this plane. The remaining planes still determine the remaining seven tetrahedra, as well as determining P_i*P_j. Each one of the seven tetrahedra is cut by four of the planes (the eight ones minus the four which contain facets of the tetrahedron, using $x_{0,1} = 0$ as was done in the first part of the proof of Theorem 2). Thus each one of the seven tetrahedra is cut into at least five parts, hence altogether into at least 35 bounded parts. P_i*P_j is cut by these eight planes into at least one bounded part (itself), thus the eight planes determine at least 36 bounded parts, contrary to the upper bound of 35 parts, established in Lemma 1. Therefore cases #4, 7, 13 and #14 are impossible, as claimed by Theorem 3.

Proof of Theorem 4: In cases #19 and #24, $x_{0,1} = 2$, $x_{1,1} = 1$ and there are altogether ten planes. Using (5), let us change the indices, if needed, such that P_1 and P_2 are the two members of F which are of the type (3, 3, 2, 0). Let H_1 be the plane, counted by $x_{1,1}$. The two members of F which have a facet on H_1 are of type which contains the number 1; therefore they are different from P_1 and P_2. Let them be P_3 and P_4. Let H_2 and H_3 be the planes which contain the free facets of P_1 and P_2, respectively. We delete the three planes H_1, H_2 and H_3. Each one of the remaining five tetrahedra is cut by three planes (the total ten, minus the deleted three and also minus the four planes containing facets of the said tetrahedron). Thus, the five remaining tetrahedra are cut into at least four bounded parts each, which together with at least one bounded part determined inside P_3*P_4 (by the seven planes: the given ten ones, less H_1, H_2 and H_3) yield at least 21 bounded parts. This contradicts Lemma 1, since seven planes can determine at most 20 bounded parts. This completes the proof of Theorem 4.

The cases, proven so far to be impossible, are marked as such in Table 1.

Chapter 3

We turn next to analyze the possible combinations of different types of tetrahedra in the remaining cases. Recall that α, β, γ and δ denote the number of members of F of type (3, 3, 2, 0), (3, 3, 1, 1), (3, 2, 2, 1) and (2, 2, 2, 2), respectively. Let r_i denote the number of facets of members of F which contain precisely i of the 36 intersections $P_j \cap P_k$ of pairs of members of F. The quantity r_i can be counted by considering the different planes (or, equivalently, the different columns of B(F)), or else by considering the number of these intersections which are contained in the facets of tetrahedra of each one of the types. Thus, an elementary double-counting yields

$$(11) \quad r_1 = \gamma + 2\beta = 2x_{1,1} + 2x_{1,2} + 3x_{1,3}$$

$$(12) \quad r_2 = \alpha + 2\gamma + 4\delta = x_{1,2} + 4x_{2,2} + 3x_{2,3}$$

$$(13) \quad r_3 = 2\alpha + 2\beta + \gamma = x_{1,3} + 2x_{2,3}$$

and equation (5) can be stated as $r_0 = \alpha = x_{0,1}$.

These equations are not independent of the previous ones, since by adding (5), (11), (12) and (13) we get $\Sigma(i+j)x_{i,j} = 4(\alpha + \beta + \gamma + \delta)$, yielding (7), since $\alpha + \beta + \gamma + \delta = |F| = 9$.

For a given decomposition (9) of K_9, r_i is also equal to the total number of maximum subgraphs of the components, subgraphs isomorphic to a $K_{1,i}$, $i = 1, 2, 3$. Another quantity which is represented by r_i is the number of minors of $B(F)$ of size $(i+1)$-by-1, containing i nonzero terms of one sign and one nonzero term of the opposite sign, and which are not included in a larger minor of the same form.

One can solve the equations (11-13, 5) in each one of the remaining cases, and find out all the possibilities for the types α, β, γ and δ. This will shed more light on the remaining cases, but it is useless in our proof, except for case #3, treated in Chapter 5. We can show that *some* of the solutions for types are impossible in each one of the remaining cases, by using arguments similar to the ones used in Chapter 5; however, we were unable to show that *all* of the solutions for types, in any one of the remaining cases, can be shown to be impossible by direct reasoning.

Chapter 4

The notion of types of members of F is translated into the notion of types of vertices of K_9 with respect to a given decomposition (9), as follows. A vertex of K_9, in a given decomposition (9), is said to be of *type* (p, q, r, s), $p \geq q \geq r \geq s$, if it is on p edges in one component $K_{i,j}$ of the decomposition, on q edges in another, etc. It follows that α, β, γ and δ are also the number of vertices of type (3, 3, 2, 0), (3, 3, 1, 1), (3, 2, 2, 1) and (2, 2, 2, 2), respectively.

property (1) of B(F) can be translated as follows:

(14) All the vertices of K_9 belong to precisely four components of the decomposition (9), except for $x_{0,1}$ of them which belong to just three components.

Let m(v), the *multiplicity* of the vertex v, denote the number of components to which a vertex v of K_9 belongs in the decomposition (9).

Thus we have

(14') $m(v) = 4$ holds for $9 - x_{0,1}$ of the vertices, while $m(v) = 3$ holds for the remaining $x_{0,1}$ of the vertices.

Consequently,

(15) $m(v) \leq 4$ holds for all the vertices of K_9 in the decomposition (9).

The next few lemmas are the main tools in Baston's work [2]. We present them here in order to make our proof as complete and self-contained as possible.

Lemma 4: If $Q = \{x \in E^d | \ Ax \leq b\}$, where A is a kxd-matrix, $b \in E^k$, and if A has a column which consists of nonnegative (or nonpositive) entries, then Q is either empty or unbounded.

This lemma is well known. It is used in the Simplex algorithm in Linear Programming, where, after getting a feasible point, if pivoting is still required but impossible to perform (in the nondegenerate case), then the conclusion is that the problem is unbounded. The proof of Lemma 4 is quite elementary: if the j-th column of A has, say, only nonnegative entries, and if Q is not empty, then take $(x_1, \dots, x_j, \dots, x_d)$ in Q. It follows that for all $z \leq x_j$, the point $(x_1, \dots, x_{j-1}, z, x_{j+1}, \dots, x_d)$ is also in Q, thus Q is unbounded.

A tetrahedron is neither empty nor unbounded, hence we get the following lemma as a particular case of Lemma 4.

Lemma 5: If a tetrahedron in E^3 is the intersection of the following four half-spaces, given by the inequalities

$$a(1)x + a(2)y + a(3)z + a(4) \geq 0$$
$$b(1)x + b(2)y + b(3)z + b(4) \geq 0$$
$$c(1)x + c(2)y + c(3)z + c(4) \geq 0$$
$$d(1)x + d(2)y + d(3)z + d(4) \geq 0,$$

then for each i, $1 \leq i \leq 3$, the set $\{a(i), b(i), c(i), d(i)\}$ contains both a positive and a negative number.

Following Baston [2], let A denote the inequality $a(1)x + a(2)y + a(3)z + a(4) \geq 0$, let B denote the inequality $b(1)x + b(2)y + b(3)z + b(4) \geq 0$, etc. We may choose any one member of F to be used as coordinates, in the sense that this member will be defined by the inequalities $x \geq 0$, $y \geq 0$, $z \geq 0$ and $1 - x - y - z \geq 0$. We denote these inequalities by the letters x, y, z and w, respectively, as Baston did. If P_i is chosen to be used for coordinates, we will denote it by an asterisk * on the i-th row and we will multiply by -1 the columns of B(F), if needed, such that the i-th row contains nonzero terms of the same sign. Above the columns where these terms appear we will write x, y, z and w (in any order we choose), as follows:

```
          x    y    z    w                       x    w    y    z
P_i*    ...1 ... 1... 1 ... 1...     or    P_i*    ...-1 ... -1... -1 ... -1...
```

If a letter, say A, appears above the j-th column of B(F), then P_i has the inequality $a(1)x + a(2)y + a(3)z + a(4) \geq 0$ as one of the four defining inequalities, in case $b_{i,j} = 1$. In case $b_{i,j} = -1$, then P_i has the opposite inequality $-a(1)x - a(2)y - a(3)z - a(4) \geq 0$ as one of the four defining inequalities. It is to be understood that if the row of the tetrahedron which defines the coordinates has -1 for all of its nonzero terms, then the corresponding inequalities are $-x \geq 0$, $-y \geq 0$, $-z \geq 0$ and $1 + x + y + z \geq 0$ (notice the difference in the last inequality!).

The next two lemmas are due to Baston.

Lemma 6: (See Lemma 4, p.5 of [2]) The Baston matrix of a neighborly family of tetrahedra contains no minors of the form

$$\begin{pmatrix} +1 & +1 & +1 \\ +1 & +1 & +1 \end{pmatrix} .$$

Proof of Lemma 6: Suppose, on the contrary, that there exists a neighborly family F* of tetrahedra, for which $B(F^*)$ contains a minor of the said form. Let Q_1 and Q_2 denote the two members of F^* which correspond to the rows of this minor. By property (2), it follows that $B(\{Q_1, Q_2\})$ is of the form

$$B(\{Q_1,Q_2\}) = \begin{array}{c} \\ Q_1 \\ Q_2 \end{array} \begin{array}{c} \begin{array}{cccc} x & y & z & w \end{array} \\ \begin{pmatrix} +1 & +1 & +1 & +1 \\ +1 & +1 & +1 & -1 \end{pmatrix} \end{array},$$

in which we have chosen Q_1 for coordinates. The row of Q_2 contradicts Lemma 5, since it is defined by the inequalities $x \geq 0, y \geq 0, z \geq 0$, and $-1 + x + y + z \geq 0$.

It follows from Lemma 6 that any two tetrahedra in a neighborly family can share the intersection of at most two half spaces, created by the planes which contain their facets.

Lemma 7: (See Lemma 5, p.6 of [2]). The Baston matrix of a neighborly family of tetrahedra contains no minors of the form

$$\begin{pmatrix} +1 & +1 \\ +1 & +1 \\ +1 & +1 \end{pmatrix}.$$

Proof of Lemma 7: Suppose, on the contrary, that there exists a neighborly family F^* of tetrahedra, for which $B(F^*)$ contains a minor of the said form. Let Q_1, Q_2 and Q_3 denote the three members of F^* which correspond to the rows of this minor. By properties (1), (2) and Lemma 6, it follows that $B(\{Q_1,Q_2,Q_3\})$ is of the form

$$B(\{Q_1,Q_2,Q_3\}) = \begin{array}{c} \\ Q_1 \\ Q_2 \\ Q_3 \end{array} \begin{array}{c} \begin{array}{ccccc} x & y & z & w & A \end{array} \\ \begin{pmatrix} +1 & +1 & +1 & +1 & 0 \\ +1 & +1 & -1 & 0 & +1 \\ +1 & +1 & 0 & -1 & -1 \end{pmatrix} \end{array},$$

in which the i-th row corresponds to Q_i, $i = 1, 2, 3$, and we have chosen Q_1 for coordinates. Let the fifth column correspond to A, i.e., one of the defining inequalities for Q_2 is

$a(1)x + a(2)y + a(3)z + a(4) \geq 0$, while one of the defining inequalities for Q_3 is $- a(1)x - a(2)y - a(3)z - a(4) \geq 0$. Thus, considering the third row, Q_3 is defined by x, y, -w and by -A, i.e., by $x \geq 0$, $y \geq 0$, $-1 + x + y + z \geq 0$ and by $- a(1)x - a(2)y - a(3)z - a(4) \geq 0$. It follows by Lemma 5 that $a(1) > 0$. Considering the second row, we get that Q_2 is defined by the inequalities $x \geq 0, y \geq 0, - z \geq 0$ and by $A = a(1)x + a(2)y + a(3)z + a(4) \geq 0$. This is a contradiction to Lemma 5, since all the coefficients of the variable x are nonnegative.

This completes the proof of Lemma 7.

To simplify the presentations of the many proofs that we are going to present, showing that certain matrices are not the Baston matrix of a neighborly family of tetrahedra, and which are similar to the one given in Lemma 7, we introduce the following code: "R.i : $f(j) \leq, \geq 0$" means that by applying Lemma 5 to the inequality which defines the tetrahedron F_i, corresponding to the i-th row of the matrix, it follows that the coefficient $f(j)$ of F is non-positive or it is non-negative. An asterisk * will mark the row which is used for coordinates, i.e., the one tetrahedron which is chosen to be defined by the inequalities x, y, z and w, as explained earlier. "Contradiction in R.i" means that a contradiction to Lemma 5 is implied when considering the inequalities which define the tetrahedron P_i. Using this code, the last part of the proof of Lemma 7 can be presented as follows.

$$\begin{array}{c} \\ 1^* \\ 2 \\ 3 \end{array} \begin{array}{c} \begin{array}{ccccc} x & y & z & w & A \end{array} \\ \begin{pmatrix} +1 & +1 & +1 & +1 & 0 \\ +1 & +1 & -1 & 0 & +1 \\ +1 & +1 & 0 & -1 & +1 \end{pmatrix} \end{array} \quad \begin{array}{l} \text{R.3: } a(1) > 0 \\ \text{contradiction in R.2.} \end{array}$$

The rows to the right of the matrix are to be understood as consecutive steps of the proof. They will include also statements like "if $f(j) < 0$", "if $f(j) > 0$", "hence $f(j) \leq 0$" or "hence $f(j) \geq 0$".

Naturally, for a larger matrix we might need more steps in getting a contradiction.

If a graph is isomorphic to $K_{2,3}$, the set of the two 3-valent vertices is called *the pair*, and the set of the three 2-valent vertices is called *the triple*.

Lemma 7 implies the following restriction on (9).

Lemma 8: In every decomposition (9) of K_9, related to a neighborly family of nine tetrahedra, the triples of every two $K_{2,3}$ graphs are never equal.

Another restriction on the decomposition (9), implied by Lemma 6, is the following one. Let u and v be any two vertices of K_9, and let m(u,v) denote the number of components $K_{A,B}$ in the decomposition (9) for which both u and v belong to A, or both belong to B. The quantity m(u,v) is equal to the number of minors of size 2x1 consisting of two equal nonzero terms which appear in rows u and v in the Baston matrix. Lemma 6 implies that $m(u,v) \leq 2$.

We state this restriction in the following lemma.

Lemma 9: For every u and v, $1 \leq u < v \leq 9$, and for every decomposition (9) of K_9, that arises from a neighborly family of nine tetrahedra, $m(u,v) \leq 2$.

Let $K_{A,B}$ denote a complete bipartite graph, isomorphic to a $K_{|A|,|B|}$, and having A and B as its bipartition of the vertex set.

Every vertex x of K_9 which is of type (3, 3, - , -), in relation to the decomposition (9), corresponds to a pair of components $K_{A,B}$ and $K_{C,D}$ for which $x \in A \cap C$ and $|B| = |D| = 3$, and vice versa. Thus we have the following.

Corollary 1: In every decomposition (9) of K_9, if $\alpha + \beta = 0$, then every two components isomorphic to a $K_{2,3}$ have disjoint pairs.

To simplify the computer searches, we use the following notations. Recall that the vertex set of K_9 is taken as $\{1, 2, ..., 9\}$. A member of the decomposition (9) of K_9, which is isomorphic to a $K_{2,3}$, having the pair $\{i_1, i_2\}$ and the triple $\{j_1, j_2, j_3\}$, will be uniquely represented by $i_1i_2{:}j_1j_2j_3$ (or simply by $i_1i_2j_1j_2j_3$), provided $i_1 < i_2$ and $j_1 < j_2 < j_3$.

Similarly, a component of the decomposition (9) of K_9, isomorphic to a $K_{2,2}$, is uniquely represented by the symbol $i_1i_2j_1j_2$, where $i_1 = \min\{i_1, i_2, j_1, j_2\}$, $j_1 < j_2$ and having the bipartition $\{i_1, i_2\}$ and $\{j_1, j_2\}$ of its vertex set. A component isomorphic to a $K_{1,3}$ will be represented by $i_1{:}j_1j_2j_3$, where $j_1 < j_2 < j_3$ and i_1 is the 3-valent vertex. A component isomorphic to a $K_{1,2}$ will be represented by $i_1{:}j_1j_2$ (or simply $i_1j_1j_2$), where i_1 is the 2-valent vertex and $j_1 < j_2$. An edge i_1j_1 $(= K_{1,1})$ will be represented by $i_1{:}j_1$, provided $i_1 < j_1$.

In this way a component of the decomposition (9) of K_9 which is isomorphic to a $K_{n,m}$ has a unique representation as a natural number having $n+m$ distinct digits. Knowing the values of n and m, we can easily deduce the edges of the component, as well as any other information needed about it.

We will denote by $K_{i,j}(t)$ the t-th component in the decomposition which is isomorphic to a

$K_{i,j}$. Without loss of generality, we will always take $K_{2,3}(1)$ to be 12:345, and we will simply write it in the form $K_{2,3}(1) = 12345$ (or $K_{2,3}(1) = 12:345$).

The idea of the proof that the remaining cases in Table 1 are impossible is the following. We produce by a computer all the possible decompositions of K_9 of the form (9), subject to all the various restrictions proven to hold; the main problem is to avoid getting isomorphic duplicates, obtained by merely permuting the vertices. Some duplications are avoided by the machine program; most of the rest are eliminated by hand, once we have a complete output. For each decomposition that we get, we present its corresponding $0, \pm 1$ matrix together with a proof (using our code) that the assumption that this matrix is the $B(F)$ of the neighborly family F of nine tetrahedra leads to a contradiction. In some cases the said decomposition contradicts one of the proven properties.

We need the following.

Lemma 10: If $K_n = \Sigma x_{i,j} K_{i,j}$ is a decomposition of K_n, such that $\Sigma x_{i,j} = n - 1$, then for every two components $K_{A,B}$ and $K_{C,D}$ of the decomposition, $A \neq C$ (and therefore also $A \neq D$, $B \neq C$ and $B \neq D$).

Proof of Lemma 10: Suppose, on the contrary, that K_n has a decomposition of the form $K_n = \Sigma x_{i,j} K_{i,j}$, having just $n - 1$ components, two of which are $K_{A,B}$ and $K_{C,D}$, and $A = C$. It follows that $B \cap D = \emptyset$, since $K_{A,B}$ and $K_{C,D}$ ($= K_{A,D}$) are edge-disjoint. Therefore $K_{A,B} \cup K_{A,D} = K_{A,B \cup D}$, implying that K_n has a decomposition into $n - 2$ complete bipartite graphs, contrary to the Graham-Pollak Theorem [7].

This completes the proof of Lemma 10.

Chapter 5

This chapter is devoted to show that case #2 of Table 1 is impossible.

Suppose the decomposition (9) of K_9 is given by

$$K_9 = K_{2,3} + 7K_{2,2} + K_{1,2} .$$

The two 3-valent vertices in $K_{2,3}$ are the only vertices of a type which is of the form (3, q, r, s), for $2 \geq q \geq r \geq s$. It follows that the types must be $\alpha = \beta = 0$, $\gamma = 2$ and $\delta = 7$.

As mentioned earlier, we let $K_{2,3} = 12{:}345$. In the given decomposition, $K_{2,3}$ and $K_{1,2}$ are the only components having vertices of odd degree; each one of them has two such vertices. It follows, therefore, that these two couples of vertices must coincide. Hence $K_{1,2} = x{:}12$, for some $x \geq 6$. We assume, without loss of generality, that $x = 6$. Thus $K_{1,2} = 6{:}12$ and $K_{2,3} + K_{1,2} = 12{:}345 + 6{:}12 = 12{:}3456$; i.e., $K_{2,3} + K_{1,2}$ is isomorphic to a $K_{2,4}$. Thus we are led to the decomposition

$$K_9 = 12{:}3456 + \bigcup_{i=1}^{7} K_{2,2}(i)$$

for which Lemma 10 is applicable. Therefore no one of the $K_{2,2}(i)$ is of the form 12:yz.

In the first part of the algorithm, we let the machine produce all the possible $K_{2,2}$ in $K_9 \setminus 12{:}3456$, i.e., all the natural numbers expressed as four distinct digits, such that when considered as a $K_{2,2}$, they do not contain the edges of 12:3456. There are, by computer search, 141 possibilities, which we denote by $H_1(i)$, $1 \le i \le 141$, where $H_1(i)$ is less than $H_1(j)$, as four digits numbers), for $i < j$.

It follows from valence considerations that the vertex 1 belongs to two of the $K_{2,2}$'s. Thus, if we insist that $K_{2,2}(i) < K_{2,2}(j)$ for all $1 \le i < j \le 7$, then $1 \in K_{2,2}(1)$ (and also that $1 \in K_{2,2}(2)$). We listed the twenty one $K_{2,2}$'s which contain the vertex 1. One of them must be $K_{2,2}(1)$. However, it suffices to consider only the four possibilities 1378, 1678, 1728 and 1789, because each one of the remaining cases can be obtained (from one of these four which appears before it) by a permutation of $\{1,...,9\}$ which preserves 12:345 and 6:12, i.e., a permutation which preserves the subsets $\{1,2\}$, $\{3,4,5\}$ and $\{7,8,9\}$.

The twenty one $K_{2,2}$'s are : *1378, 1379, 1389, 1478, 1479, 1489, 1578, 1579, 1589, *1678, 1679, 1689, *1728, 1729, *1789, 1827, 1829, 1879, 1927, 1928 and 1978.

$K_{2,2}(1) \ne 1789$, because the vertex 1 has to appear also in $K_{2,2}(2)$, but the given list does not contain any $K_{2,2}$, appearing *after* 1789, and which is edge-disjoint from 1789. For the same reason $K_{2,2}(1) \ne 1728$. Thus it suffices to assume that $K_{2,2}(1)$ is either 1378 or 1678; i.e., $K_{2,2}(1)$ is either $H_1(1)$ or it is $H_1(10)$; in other words: $K_{2,2}(1) = H_1(j)$ for $j = 1$ or 10.

Let H^*_1 denote the list $(H_1(1),..., H_1(141))$. Let H^*_2 denote the sublist of H^*_1, containing all those $H_1(i)$, $j < i \le 141$, which are edge disjoint from $12{:}3456 + H_1(j)$, for the cases $j = 1$ or $j = 10$. $H^*_2 = (H_2(1), H_2(2),...)$, where H^*_2 depends on j. We determined (by hand) those indices i_2 for which $H_2(i_2)$ is a candidate for $K_{2,2}(2)$, assuring that those untreated possibilities will lead to isomorphic duplicates (of at least one $K_{2,2}(2)$ which *is* treated). For every choice of i_2, we let the machine produce the subset of H^*_2, consisting of all those $H_2(i)$, $i > i_2$, which are edge disjoint from $H_2(i_2)$. Let H^*_3 denote the resulting sublist of H^*_2, where $H^*_3 = (H_3(1), H_3(2), \ldots)$; H^*_3 depends on j and i_2. For each member $H_3(i_3)$ in H^*_3, we let the machine search for all those members $H_3(i)$ in H^*_3, $i > i_3$, which are edge disjoint from $H_3(i_3)$. These are the candidates for $K_{2,2}(4)$, assuming we are trying $12{:}3456 + H_1(j) + H_2(i_2) + H_3(i_3)$. The result is a subset $H^*_4 = (H_4(1), H_4(2), \ldots)$, which depends on j, i_2 and i_3. In this way the machine produces a nested collection of lists, $H^*_2 \supset H^*_3 \supset \ldots \supset H^*_7$. If $H^*_7 \neq \emptyset$, then we get a decomposition of K_9 into $12{:}345 + 6{:}12 + H_2(i_2) + \ldots + H_6(i_6)$ plus each member (separately) of H^*_7. If $i_k < |H^*_k|$ and $i_j = |H^*_j|$ for all $j > k$, then i_k is replaced by $i_k{+}1$, and the machine redefines H^*_{k+1}, and so on. If during the process a sublist H^*_k is found to be empty, then $H_{k-1}(i_{k-1})$ is replaced by $H_{k-1}(i_{k-1} + 1)$, unless $i_{k-1} = |H^*_{k-1}|$, in which case $H_{k-2}(i_{k-2})$ is replaced by $H_{k-2}(i_{k-2} + 1)$, and so on.

Throughout the search, the machine discards any beginning of a decomposition which violates the proven restrictions stated in Lemma 10 and in Lemma 9.

The machine found out a total of ninety decompositions of K_9, satisfying the established restrictions, and which are of the form $12{:}345 + 6{:}12$ plus either $13{:}78$ or $16{:}78$ plus six other $K_{2,2}$'s. Seventy of them were shown by hand to be isomorphic duplicates of one of the remaining twenty. Each one of these twenty decompositions is presented, together with the corresponding 9x9-matrix and a proof, using our code, that the matrix cannot be the Baston matrix $B(F)$ of a neighborly family F of nine tetrahedra. Each one of these matrices leads to a contradiction, as follows (terms which do not appear are zero).

5.1

		x	D	A	E	B	y	z	C	w
12:345										
6:12	1	1	1	-1	1					
13:78	2	1	1		-1	1				
17:29	3	-1		-1			-1	1		
24:89	4	-1				1	1		1	
35:46	5*	-1					-1	-1		-1
38:59	6		-1				1		-1	-1
48:67	7			1	1				-1	1
56:79	8			1		-1		1	1	
	9				-1	-1		-1		1

R.3: $a(3) > 0$
if $b(3) > 0$
R.4: $c(3) < 0$
R.6: $d(3) > 0$
R.9: $e(3) < 0$
contradiction in R.2;
hence $b(3) \le 0$
R.8: $c(3) < 0$
contradiction in R.4.

The proof of case 5.1 is to be read as follows: the decomposition of K_9 is given as 12:345 + 6:12 + 13:78 + 17:29 + 24:89 + 35:46 + 38:59 + 48:67 + 56:79. The corresponding B(F) matrix is presented, in which the columns were multiplied by -1, if needed, so that the four nonzero terms in the fifth row are all of the same sign (negative in this case). The fifth tetrahedron is chosen to be defined by the inequalities -x (the first column), -y (the sixth column), -z (the seventh column) and -w (the ninth column). The remaining columns are denoted by A, B, C etc, usually in the order they appear in the proof (at the right side of the matrix). Thus, in the third row we can conclude about one coefficient in the equation of the plane, presented in the third column; let A present the inequality of the third column. We conclude that $a(3) > 0$, which is what is meant by "R.3: $a(3) > 0$", and so on.

5.2

12:345
6:12
13:78
17:29
24:89
35:49
37:56
48:67
59:68

	x	D	A	E	B	y	z	C	w
1	1	1	1	1					
2	1	1		-1	1				
3	-1		1			-1	1		
4	-1				1	1		1	
5 *	-1					-1	-1		-1
6		-1					-1	-1	1
7			-1	1			1	-1	
8			-1		-1			1	1
9				-1	-1	1			-1

R.3: $a(3) < 0$
if $b(3) < 0$
R.4: $c(3) > 0$
R.6: $d(3) < 0$
R.1: $e(3) > 0$
contradiction in R.2;
hence $b(3) \geq 0$
R.4: $c(3) < 0$
R.7: $e(3) < 0$
R.1: $d(3) > 0$
contradiction in R.2.

5.3

12:345
6:12
13:78
17:29
24:89
35:49
38:56
45:67
68:79

	x	D	B	E	A	y	z	w	C
1	1	1	1	1					
2	1	1		-1	1				
3	-1		1			-1	1		
4	-1				1	1		-1	
5 *	-1					-1	-1	-1	
6		-1					-1	1	1
7			-1	1				1	-1
8			-1		-1		1		1
9				-1	-1	1			-1

R.4: $a(2) < 0$
R.3: $b(2) > 0$
if $d(2) < 0$
R.2: $e(2) < 0$
R.7: $c(2) < 0$
contradiction in R.9;
hence $d(2) \geq 0$
R.1: $e(2) < 0$
R.6: $c(2) > 0$
contradiction in R.7.

5.4

12:345
6:12
13:78
17:29
24:89
35:69
37:45
48:56
68:79

	x	C	B	D	A	y	z	w	E
1	1	1	1	1					
2	1	1		-1	1				
3	-1		1			-1	1		
4	-1				1		-1	1	
5 *	-1					-1	-1	-1	
6		-1				1		-1	1
7			-1	1			1		-1
8			-1		-1			1	1
9				-1	-1	1			-1

R.4: $a(1) > 0$
R.3: $b(1) > 0$
if $d(1) < 0$
R.2: $c(1) < 0$
R.6: $e(1) < 0$
contradiction in R.8;
hence $d(1) \geq 0$
R.1: $c(1) < 0$
R.6: $e(1) < 0$
contradiction in R.8.

5.5

		x	B	D	C	A	y	E	w	z
12:345										
6:12	1	1	1	1	1					
13:78	2	1	1		-1	1				
17:29	3	-1		1			1	1		
24:89	4	-1				1	-1		1	
36:45	5 *	-1					-1		-1	-1
38:69	6		-1				1	-1		-1
48:57	7			-1	1				-1	1
56:79	8			-1		-1		1	1	
	9				-1	-1		-1		1

R.4: a(2) > 0
if b(2) > 0
R.2: c(2) > 0
R.1: d(2) < 0
contradiction in R.7;
hence b(2) ≤ 0
R.6: e(2) > 0
R.9: c(2) < 0
R.2: b(2) < 0
R.1: d(2) > 0
contradiction in R.3.

5.6

		x	B	E	D	C	A	y	z	w
12:345										
6:12	1	1	1	1	1					
13:78	2	1	1		-1	1				
17:29	3	-1		1			1	1		
24:89	4	-1				1	-1		1	
36:49	5 *	-1						-1	-1	-1
37:56	6		-1				1	-1		1
48:57	7			-1	1			1	-1	
58:69	8			-1		-1			1	-1
	9				-1	-1	-1			1

If a(3) < 0
R.6: b(3) < 0
R.4: c(3) < 0
R.2: d(3) < 0
R.1: e(3) > 0
contradiction in R.7;
hence a(3) ≥ 0
R.3: e(3) < 0
R.8: c(3) > 0
R.9: d(3) < 0
R.1: b(3) > 0 , contradiction in R.2.

5.7

		x	B	C	E	D	A	y	w	z
12:345										
6:12	1	1	1	1	1					
13:78	2	1	1		-1	1				
17:29	3	-1		1			1	1		
24:89	4	-1				1	-1		1	
36:49	5 *	-1						-1	-1	-1
38:56	6		-1				1	-1	1	
46:57	7			-1	1				-1	1
58:79	8			-1		-1		1		-1
	9				-1	-1	-1			1

If a(3) < 0
R.6: b(3) < 0
R.3: c(3) > 0
R.8: d(3) < 0
R.9: e(3) > 0
contradiction in R.2;
hence a(3) ≥ 0
R.3: c(3) < 0
R.4: d(3) > 0
R.7: e(3) < 0
R.1: b(3) > 0 , contradiction in R.2.

5.8

12:345
6:12
13:78
17:29
24:89
36:59
37:46
48:56
58:79

	x	D	C	A	B	w	E	y	z
1	1	1	1	1					
2	1	1		-1	1				
3	-1		1			1	1		
4	-1				1		-1	1	
5*	-1					-1		-1	-1
6		-1				1	-1	-1	
7			-1	1			1		1
8			-1		-1			1	-1
9				-1	-1	-1			1

If $b(3) < 0$
R.9: $a(3) > 0$
R.2: $d(3) > 0$
R.1: $c(3) < 0$
R.3: $e(3) > 0$
contradiction in R.4;
hence $b(3) \geq 0$
R.4: $e(3) > 0$
R.6: $d(3) < 0$
R.8: $c(3) < 0$
R.7: $a(3) < 0$, contradiction in R.1.

5.9

12:345
6:12
13:78
17:29
24:89
37:45
38:69
49:56
57:68

	x	B	D	C	A	y	E	z	w
1	1	1	1	1					
2	1	1		-1	1				
3	-1		1			1	1		
4	-1				1	-1		1	
5*	-1					-1		-1	-1
6		-1					-1	-1	1
7			-1	1		1			-1
8			-1		-1		1		1
9				-1	-1		-1	1	

R.4: $a(2) > 0$
if $e(2) > 0$
R.3: $d(2) < 0$
R.9: $c(2) < 0$
R.6: $b(2) < 0$
contradiction in R.1;
hence $e(2) \leq 0$
R.8: $d(2) < 0$
R.7: $c(2) < 0$
R.1: $b(2) > 0$
contradiction in R.2

5.10

12:345
6:12
13:78
17:29
24:89
37:46
38:59
49:56
58:67

	x	B	C	E	w	y	A	z	D
1	1	1	1	1					
2	1	1		-1	-1				
3	-1		1			1	1		
4*	-1				-1	-1		-1	
5	-1						-1	1	-1
6		-1				-1		1	1
7			-1	1		1			1
8			-1		1		1		-1
9				-1	1		-1	-1	

If $a(2) > 0$
R.9: $e(2) < 0$
R.5: $d(2) < 0$
R.6: $b(2) < 0$
R.3: $c(2) < 0$
contradiction in R.1;
hence $a(2) \leq 0$
R.5: $d(2) > 0$
R.8: $c(2) < 0$
R.7: $e(2) < 0$, R.2: $b(2) < 0$, contradiction in R.1.

5.11

		x	D	B	C	E	w	A	y	z
12:345										
6:12	1	1	1	1	1					
13:78	2	1	1		-1	1				
17:29	3	-1		1			1	1		
26:89	4	-1					1	-1	-1	
34:56	5 *	-1					-1		-1	-1
38:49	6		-1			1	-1			1
45:79	7			-1	1				1	-1
57:68	8			-1		-1		1		1
	9				-1	-1		-1	1	

R.4: a(3) < 0
R.3: b(3) > 0
R.7: c(3) > 0
R.1: d(3) < 0
R.2: e(3) > 0
contradiction in R.6.

5.12

		x	E	B	D	C	w	A	y	z
12:345										
6:12	1	1	1	1	1					
13:78	2	1	1		-1	1				
17:29	3	-1		1			1	1		
26:89	4	-1					1	-1	-1	
34:59	5 *	-1					-1		-1	-1
37:46	6		-1			1		-1	1	
45:68	7			-1	1			1		1
58:79	8			-1		-1			1	-1
	9				-1	-1	-1			1

R.4: a(3) < 0
R.3: b(3) > 0
R.8: c(3) < 0
R.9: d(3) > 0
R.2: e(3) > 0
contradiction in R.1

5.13

		x	D	B	C	E	A	y	w	z
12:345										
6:12	1	1	1	1	1					
13:78	2	1	1		-1	1				
17:29	3	-1		1			1	1		
26:89	4	-1					1	-1	1	
34:69	5 *	-1						-1	-1	-1
37:45	6		-1			1	-1			1
49:58	7			-1	1			1		-1
57:68	8			-1		-1			-1	1
	9				-1	-1	-1		1	

R.4: a(2) > 0
R.3: b(2) < 0
R.7: c(2) < 0
R.1: d(2) > 0
R.2: e(2) < 0
contradiction in R.6.

5.14

12:345
6:12
13:78
17:29
26:89
34:69
38:45
46:57
58:79

	x	E	B	C	D	A	y	w	z
1	1	1	1	1					
2	1	1		-1	1				
3	-1		1			1	1		
4	-1					1	-1	1	
5 *	-1						-1	-1	-1
6		-1			1	-1		1	
7			-1	1				-1	1
8			-1		-1		1		-1
9				-1	-1	-1			1

R.4: $a(2) > 0$
R.3: $b(2) < 0$
R.7: $c(2) < 0$
R.8: $d(2) > 0$
R.2: $e(2) < 0$
contradiction in R.1.

5.15

12:345
6:12
13:78
17:29
26:89
35:46
38:59
45:79
47:68

	x	D	B	C	E	w	A	y	z
1	1	1	1	1					
2	1	1		-1	1				
3	-1		1			1	1		
4 *	-1					-1		-1	-1
5	-1					1	-1	-1	
6		-1			1	-1			1
7			-1	1				1	-1
8			-1		-1		1		1
9				-1	-1		-1	1	

R.5: $a(3) < 0$
R.3: $b(3) > 0$
R.7: $c(3) > 0$
R.1: $d(3) < 0$
R.2: $e(3) > 0$
contradiction in R.6.

5.16

12:345
6:12
13:78
17:29
26:89
35:49
37:56
45:68
48:79

	x	D	B	C	E	w	A	y	z
1	1	1	1	1					
2	1	1		-1	1				
3	-1		1			1	1		
4 *	-1					-1		-1	-1
5	-1					1	-1	-1	
6		-1			1		-1	1	
7			-1	1			1		1
8			-1		-1			1	-1
9				-1	-1	-1			1

R.5: $a(2) < 0$
R.3: $b(2) > 0$
R.7: $c(2) > 0$
R.1: $d(2) < 0$
R.2: $e(2) > 0$
contradiction in R.6.

5.17

		x	D	B	C	E	A	y	z	w
12:345										
6:12	1	1	1	1	1					
13:78	2	1	1		-1	1				
17:29	3	-1		1			1	1		
26:89	4*	-1						-1	-1	-1
35:69	5	-1					1	-1		1
37:45	6		-1			1	-1		1	
47:68	7			-1	1			1	-1	
48:59	8			-1		-1			1	-1
	9				-1	-1	-1			1

R.5: a(2) > 0
R.3: b(2) < 0
R.7: c(2) < 0
R.1: d(2) > 0
R.2: e(2) < 0
contradiction in R.6.

5.18

		x	D	B	C	E	A	y	w	z
12:345										
6:12	1	1	1	1	1					
13:78	2	1	1		-1	1				
17:29	3	-1		1			1	1		
26:89	4*	-1						-1	-1	-1
35:69	5	-1					1	-1	1	
38:45	6		-1			1	-1		1	
47:56	7			-1	1				-1	1
48:79	8			-1		-1		1		-1
	9				-1	-1	-1			1

R.5: a(2) > 0
R.3: b(2) < 0
R.7: c(2) < 0
R.1: d(2) > 0
R.2: e(2) < 0
contradiction in R.6.

5.19

		x	D	B	C	E	A	y	z	w
12:345										
6:12	1	1	1	1	1					
13:78	2	1	1		-1	1				
17:29	3	-1		1			1	1		
26:89	4	-1					-1		-1	1
37:46	5*	-1						-1	-1	-1
38:59	6		-1			1	-1		1	
45:69	7			-1	1		1			1
47:58	8			-1		-1		1		-1
	9				-1	-1		-1	1	

R.4: a(3) < 0
R.3: b(3) > 0
R.7: c(3) > 0
R.1: d(3) < 0
R.2: e(3) > 0
contradiction in R.6.

5.20

		x	E	B	D	C	y	A	z	w
12:345	1	1	1	1	1					
6:12	2	1	1		-1	1				
13:78	3	-1		1			1	1		
17:29	4	-1						-1	-1	1
26:89	5 *	-1					-1		-1	-1
37:56	6		-1			1	-1		1	
38:49	7			-1	1		1			-1
45:69	8			-1		-1		1		1
48:57	9				-1	-1		-1	1	

R.4: $a(3) < 0$
R.3: $b(3) > 0$
R.8: $c(3) < 0$
R.9: $d(3) > 0$
R.1: $e(3) < 0$
contradiction in R.2.

These are all the twenty possibilities in case #2 of Table 1, hence case #2 is impossible.

Chapter 6

In this chapter we treat case #3 of Table 1, concerning the following decomposition of K_9:

$$K_9 = 2K_{2,3} + 5K_{2,2} + 2K_{1,2}.$$

By solving the equations (5, 11 - 13) we get that the types α, β, γ and δ are such that $(\alpha, \beta, \gamma, \delta)$ is one of $(0, 2, 0, 7)$, $(0, 1, 2, 6)$ or $(0, 0, 4, 5)$.

To see that $(\alpha, \beta, \gamma, \delta)$ differs from $(0, 2, 0, 7)$, we extend Baston's observation [2, p.179] on the connection between $x_{2,2}$ and δ, as follows. Every pair of tetrahedra of type $(2, 2, 2, 2)$ in F meet in a plane which is counted by $x_{2,2}$; there are $\delta(\delta-1)/2$ such pairs, and any one such plane can deal with at most four such pairs. It follows that

$$x_{2,2} \geq (1/4)\delta(\delta-1)/2 = \delta(\delta-1)/8 .$$

In particular, $x_{2,2} = 5$ implies that $\delta \leq 6$, thus $(\alpha, \beta, \gamma, \delta)$ differs from $(0, 2, 0, 7)$.

Suppose that $(\alpha, \beta, \gamma, \delta) = (0, 1, 2, 6)$. Let 1 be the unique vertex of type $(3, 3, 1, 1)$, and let r and s denote the two vertices of type $(3, 2, 2, 1)$. The remaining vertices are of type $(2, 2, 2, 2)$. The two $K_{2,3}$'s are of the form 1r: ... and 1s: ... . The only vertices of valence one in a component of the decomposition are the vertex 1 (twice) and the vertices r and s (one each), while the only components having a 1-valent vertex are the two $K_{1,2}$'s. It follows that the two $K_{1,2}$'s are x:1r and y:1s, for some x and y. Thus 1r: ... + x:1r and 1s: ... + y:1s are both isomorphic to a $K_{2,4}$, implying that K_9 has a decomposition into just *seven* complete bipartite graphs, in contradiction with the Graham-Pollak Theorem [7].

Thus it follows that in the aforesaid decomposition of K_9, $(\alpha, \beta, \gamma, \delta) = (0, 0, 4, 5)$. Hence, by Corollary 1, the two pairs in the $K_{2,3}$'s are disjoint.

We choose the notations such that $K_{2,3}(1)$ = 12:345. There are twelve possibilities for $K_{2,3}(2)$, assuming we do not distinguish between 1 and 2, among 3, 4 and 5 and among the rest of the vertices. We emphasize those vertices of $K_{2,3}(1)$ which appear also in $K_{2,3}(2)$ and denote these twelve possibilities of $K_{2,3}(1) + K_{2,3}(2)$ by 6*i, $1 \leq i \leq 12$, as follows.

6*1: $12{:}\underline{345} + \underline{34}{:}\underline{5}67$ **6*2:** $12{:}\underline{34}5 + \underline{34}{:}678$ **6*3:** $12{:}\underline{345} + \underline{3}6{:}\underline{45}7$

6*4: $12{:}\underline{34}5 + \underline{3}6{:}\underline{4}78$ **6*5:** $12{:}\underline{3}45 + \underline{3}6{:}789$ **6*6:** $\underline{12}{:}\underline{3}45 + 67{:}\underline{123}$

6*7: $\underline{12}{:}345 + 67{:}\underline{12}8$ **6*8:** $\underline{1}2{:}\underline{34}5 + 67{:}\underline{134}$ **6*9:** $\underline{1}2{:}\underline{3}45 + 67{:}\underline{13}8$

6*10: $\underline{1}2{:}345 + 67{:}\underline{1}89$ **6*11:** $12{:}\underline{34}5 + 67{:}\underline{34}8$ **6*12:** $12{:}\underline{3}45 + 67{:}\underline{3}89$.

Checking the list and the vertices in common to the two $K_{2,3}$'s reveals that 6*6 is combinatorially isomorphic to 6*1, and so are the following pairs: 6*7 and 6*2, 6*8 and 6*3,

6*9 and 6*4, 6*10 and 6*5. We state this as a corollary.

Corollary 2: If the components of the decomposition (9) of K_9 have disjoint pairs, then there is no loss of generality in assuming that they are as in cases 6*1, 6*2, 6*3, 6*4, 6*5, 6*11 and 6*12.

The four vertices appearing in the pairs in $K_{2,3}(1)$ and $K_{2,3}(2)$ are also the four 1-valent vertices in the two components $2K_{1,2}$, because the vertices are either of type (3, 2, 2, 1) or (2, 2, 2, 2). This is helpful in showing that 6*5 and 6*12 are impossible, as follows.

In case 6*5, $K_{2,3}(1)$ = 12:345 and $K_{2,3}(2)$ = 36:789. The two components $K_{1,2}(1)$ and $K_{1,2}(2)$ are either x:12 + y:36 or else they are x:13 + y:26, for some two vertices x and y of K_9 (the third combination, that of x:16 + y:23, is obtainable from x:13 + y:26 by interchanging 1 and 2 as well as x and y, which preserves 12:345 + 36:789).

If $K_{1,2}$ (1) = x:12 and $K_{1,2}(2)$ = y:36, then $K_{2,3}(1) + K_{1,2}(1)$ = 12:345 + x:12 forms a $K_{2,4}$, and so does $K_{2,3}(2) + K_{1,2}(2)$ = 36:789 + y:36. This implies that K_9 has a decomposition into just *seven* complete bipartite graphs, contrary to the Graham-Pollak Theorem [7].

If $K_{1,2}$ (1) = x:13 and $K_{1,2}(2)$ = y:26, then the edge-disjointness of 12:345, 36:789 and y:26 implies that y = 1. However, the edge-disjointness of 12:345, 36:789, 1:26 and x:13 leaves no choices for x. Therefore 6*5 is impossible.

In case 6*12, $K_{2,3}(1)$ = 12:345 and $K_{2,3}(2)$ = 67:389. It follows that for some two vertices x and y of K_9, $2K_{1,2} = x{:}12 + y{:}67$, or else $2K_{1,2} = x{:}16 + y{:}27$. The first possibility leads, as in the previous case, to a contradiction to the Graham-Pollak Theorem [7]. In the second possibility, the edge-disjointness of 12:345, 67:389 and x:16 implies that x equals to either 2 or 7. The edge-disjointness of 12:345, 67:389 and y:27 implies that y equals to either 1 or 6. However, no one of the four possible values for both x and y leads to an edge-disjoint collection of the form 12:345, 67:389, x:16 and y:27. Therefore 6*12 is impossible.

In case 6*2, $K_{2,3}(1) = 12{:}345$ and $K_{2,3}(2) = 34{:}678$. It follows that $2K_{1,2} = x{:}12 + y{:}34$ or else $2K_{1,2} = x{:}13 + y{:}24$. The first possibility leads to a contradiction to the Graham-Pollak Theorem. In the second possibility, it follows by edge-disjointness arguments that $x = y = 9$; thus $2K_{1,2} = 9{:}13 + 9{:}24$. This can be replaced by $2K_{1,2} = 9{:}12 + 9{:}34$, which when combined with 12:345 and 34:678 leads to a contradiction to the Graham-Pollak Theorem [7]. Thus case 6*2 is also impossible.

In each one of the remaining cases 6*1, 6*3, 6*4 and 6*11, having determined the two $K_{2,3}$'s, we let the machine find all the possibilities for $K_9\backslash(2K_{2,3} + 5K_{2,2})$, in the following way. We first list all the possible $K_{2,2}$ in $K_9\backslash 2K_{2,3}$. We then determine which are the candidates for $K_{2,2}(1)$, making sure that those untreated would lead to isomorphic duplicates of the ones treated. For each such a $K_{2,2}(1)$ we use again the idea of nested collections of lists for the remaining four $K_{2,2}$'s.

Whenever a $K_9\backslash(2K_{2,3} + 5K_{2,2})$ is found, subject to the restriction of Lemma 9, we let the machine check if this graph on four edges contains an isolated edge, i.e., an edge which is a connected component of the graph. If an isolated edge is found, then this possibility for $K_9\backslash(2K_{2,3} + 5K_{2,2})$ is not the union of two $K_{1,2}$'s, the case is dropped, and the search continues. All the cases in which $K_9 \setminus (2K_{2,3} + 5K_{2,2})$ has no isolated edges were recorded. We manually discarded isomorphic duplicates. The twenty six cases which have been obtained in this way are presented, together with the corresponding matrix and a coded proof that each one of them leads to a contradiction, as follows.

6.1

		B	x	w	A	y	D	z	C	E
12:345	1	1		-1	1				1	
34:567	2	1			-1	-1				-1
15:67	3	-1	-1				1			-1
16:28	4	-1	-1				-1		1	
26:79	5	-1	1	-1				1		
38:49	6		1	1	1	-1				
57:89	7*		1	1		1		1		
9:14	8				-1		1	-1		1
8:23	9					1	-1	-1	-1	

R.6: $a(3) > 0$
R.5: $b(3) > 0$
R.1: $c(3) < 0$
R.4: $d(3) < 0$
R.3: $e(3) < 0$
contradiction in R.8.

6.2

		B	x	A	y	z	D	w	C	E
12:345	1	1		1	1				-1	
34:567	2	1			-1	1				-1
15:68	3	-1	1				1			-1
16:27	4	-1	1				-1		-1	
25:79	5	-1	-1	1		1				
38:49	6		-1	-1	1			-1		
67:89	7*		-1		-1	-1		-1		
9:14	8			-1			1	1		1
8:23	9					-1	-1	1	1	

R.6: $a(2) > 0$
R.5: $b(2) > 0$
R.1: $c(2) > 0$
R.4: $d(2) < 0$
R.3: $e(2) < 0$
contradiction in R.8.

6.3

		B	x	A	y	z	D	w	E	C
12:345	1	1		1	1				-1	
34:567	2	1			-1	1				-1
15:79	3	-1	1				1		-1	
18:26	4	-1	1				-1			-1
29:67	5	-1	-1	1				1		
38:49	6*		-1		-1	-1		-1		
57:68	7		-1	-1		-1		1		
8:13	8				1		1	-1	1	
9:24	9			-1		1	-1			1

R.7: $a(1) < 0$
R.5: $b(1) < 0$
R.2: $c(1) < 0$
R.4: $d(1) > 0$
R.1: $e(1) < 0$
contradiction in R.3.

6.4

		A	x	B	y	z	D	w		C
12:345	1	1		1	1					-1
34:567	2	1		-1		1			-1	
16:28	3	-1	1				1		-1	
19:67	4	-1	1				-1			-1
25:79	5	-1	-1			1		-1		
38:49	6		-1	1	-1			1		
57:68	7*		-1		-1	-1		-1		
8:23	8			-1			1	1	1	
9:14	9				1	-1	-1			1

R.5: $a(2) > 0$
R.6: $b(2) > 0$
R.1: $c(2) > 0$
R.4: $d(2) < 0$
contradiction in R.9.

6.5

		x	y	A	B	z	D	w		C
12:345	1	-1		1	1				1	
34:567	2	-1			-1	1				-1
16:79	3	1	-1				1		1	
18:26	4	1	-1				-1			-1
25:67	5*	1	1			1		1		
38:49	6		1	1	-1	-1				
57:89	7		1	-1		-1		1		
8:13	8				1		1	-1	-1	
9:24	9			-1			-1	-1		1

R.7: $a(3) < 0$
R.6: $b(3) < 0$
R.2: $c(3) > 0$
R.4: $d(3) < 0$
contradiction in R.9.

6.6

12:345
36:457
13:68
14:79
29:78
47:58
56:89
2:16
9:23

	x	y	C	A	B	z	w	D	E
1	1		1	1				-1	
2	1				1			1	-1
3	-1	1	1						-1
4	-1	-1		1		1			
5*	-1	-1				-1	-1		
6		1	-1				-1	-1	
7		-1		-1	-1	1			
8			-1		-1	-1	1		
9				-1	1		1		1

R.4: $a(3) < 0$
R.7: $b(3) > 0$
R.8: $c(3) < 0$
R.6: $d(3) > 0$
R.2: $e(3) > 0$
contradiction in R.3.

6.7

12:345
36:457
13:68
15:79
24:78
48:59
67:89
2:16
9:23

	x	y	C	z	A	w	B	D	E
1	1		1	-1				-1	
2	1				-1			1	-1
3	-1	1	1						-1
4	-1	-1			-1	1			
5*	-1	-1		-1		-1			
6		1	-1				1	-1	
7		-1		1	1		1		
8			-1		1	1	-1		
9				1		-1	-1		1

R.4: $a(1) < 0$
R.7: $b(1) > 0$
R.8: $c(1) < 0$
R.6: $d(1) > 0$
R.3: $e(1) < 0$
contradiction in R.2.

6.8

12:345
36:457
13:68
18:79
24:78
47:59
56:89
2:16
9:23

	x	y	C	B	A	z	w	D	
1	1		1	1				-1	
2	1				1			1	-1
3	-1	1	1						-1
4	-1	-1			1	1			
5*	-1	-1				-1	-1		
6		1	-1				-1	-1	
7		-1		-1	-1	1			
8			-1	1	-1		1		
9				-1		-1	1		1

R.4: $a(2) > 0$
R.7: $b(2) < 0$
R.8: $c(2) < 0$
R.1: $d(2) < 0$
contradiction in R.6.

6.9

		A	x	D	w	y	B	z		C
12:345	1	1		-1	1				1	
36:457	2	1				1			-1	-1
13:69	3	-1	1	-1						-1
15:78	4	-1	-1				1	1		
28:79	5	-1	-1		1			-1		
46:89	6		1	1			1		1	
49:57	7 *		-1		-1	-1		-1		
2:16	8				-1	1	-1			1
8:23	9			1		-1	-1	1		

R.5: $a(3) < 0$

R.4: $b(3) < 0$

R.8: $c(3) < 0$

R.3: $d(3) > 0$

contradiction in R.9.

6.10

		x	y	z	D	A	B	w	C	
12:345	1	1		-1	1				-1	
36:457	2	1			-1	-1				-1
14:78	3	-1	1				1		-1	
16:29	4 *	-1	-1	-1				-1		
25:79	5	-1	-1			-1		1		
37:89	6		1		1				1	-1
48:59	7		-1	1		1	1			
6:13	8			1			-1	-1		1
8:26	9				-1	1	-1	1		

R.5: $a(2) < 0$

R.7: $b(2) > 0$

R.3: $c(2) > 0$

R.1: $d(2) > 0$

contradiction in R.9.

6.11

		x	y	z	C	E	A	w	B	D
12:345	1	1		-1	1				-1	
36:457	2	1			-1	1				-1
14:78	3	-1	1				1		-1	
16:29	4 *	-1	-1	-1				-1		
28:79	5	-1	-1				1	1		
35:89	6		1		1				1	-1
47:59	7		-1	1		-1		-1		
6:13	8			1		1	-1			1
8:26	9				-1	-1	-1	1		

R.5: $a(2) > 0$

R.3: $b(2) > 0$

R.1: $c(2) > 0$

R.6: $d(2) > 0$

R.2: $e(2) > 0$

contradiction in R.9.

6.12

		x	y	A	D	z	B	w	C	
12:345	1	1		-1	1					-1
36:457	2	1			-1	-1			-1	
14:78	3	-1	1				1		-1	
18:26	4	-1	-1	-1				1		
25:79	5 *	-1	-1			-1		-1		
37:89	6		1		-1				1	-1
48:59	7		-1	1		1	1			
6:23	8			1	1		-1	1		
9:16	9					1	-1	-1		1

R.4: $a(2) < 0$
R.7: $b(2) > 0$
R.3: $c(2) > 0$
R.2: $d(2) < 0$
contradiction in R.8.

6.13

		x	y	A	D	B	z	w		C
12:345	1	1		1	1					-1
36:457	2	1			-1	1			-1	
14:78	3	-1	1				-1		-1	
18:26	4	-1	-1	1				1		
28:79	5 *	-1	-1				-1	-1		
35:89	6		1		-1				1	-1
47:59	7		-1	-1		-1		1		
6:23	8			-1	1	1	1			
9:16	9					-1	1	-1		1

R.4: $a(1) > 0$
R.7: $b(1) < 0$
R.9: $c(1) < 0$
R.8: $d(1) > 0$
contradiction in R.1.

6.14

		x	y	A	C	B	z	w	D	C
12:345	1	1		1	1				-1	
36:457	2	1			-1	1				-1
14:78	3	-1	1				-1		-1	
18:29	4	-1	-1	1				1		
28:67	5 *	-1	-1				-1	-1		
35:89	6		1			-1			1	-1
47:59	7		-1	-1		-1		1		
6:13	8			-1	1	1	1			
9:26	9				-1		1	-1		1

R.4: $a(2) > 0$
R.7: $b(2) < 0$
R.8: $c(2) > 0$
R.3: $d(2) > 0$
R.2: $e(2) < 0$
contradiction in R.6.

6.15

		x	y	A	B	z	E	w	D	C
12:345	1	1		1	1					-1
36:457	2	1				-1			-1	1
14:89	3	-1	1				1		-1	
19:67	4	-1	-1	1				1		
25:79	5 *	-1	-1			-1		-1		
38:69	6		1		-1		-1			-1
48:57	7		-1		-1	1		-1		
8:23	8			-1			1	1	1	
2:16	9			-1	1	1	-1			

R.4: a(1) > 0
R.7: b(1) > 0
R.1: c(1) > 0
R.2: d(1) > 0
R.3: e(1) > 0
contradiction in R.6.

6.16

		x	y	C	B	A	z	w		D
12:345	1	1		1	1				-1	
36:457	2	1		-1		1				-1
16:28	3	-1	1				-1		-1	
18:79	4	-1	-1			1		1		
24:78	5 *	-1	-1				-1	-1		
35:89	6		1	1					1	-1
47:59	7		-1		-1	-1		1		
6:13	8			-1	1	-1	1			
9:26	9				-1		1	-1		1

R.4: a(1) > 0
R.7: b(1) < 0
R.8: c(1) < 0
R.2: d(1) > 0
contradiction in R.9.

6.17

		x	y	C	E	z	A	w	D	B
12:345	1	1		-1	1				-1	
36:457	2	1		1		-1				-1
18:26	3	-1	1				1			-1
19:78	4 *	-1	-1			-1		-1		
24:79	5	-1	-1				1	1		
35:89	6		1	1					-1	1
47:58	7		-1		-1	1		-1		
9:16	8			-1	-1		-1	1		
6:23	9				1	1	-1		1	

R.5: a(2) > 0
R.3: b(2) > 0
R.2: c(2) > 0
R.6: d(2) > 0
R.1: e(2) > 0
contradiction in R.8.

6.18

12:345
36:457
18:79
19:26
24:78
35:89
47:59
6:23
8:16

	x	y	E	C	A	z	w	B	D
1	1		1	1					-1
2	1			-1	1			-1	
3	-1	1				-1		-1	
4	-1	-1			1		1		
5 *	-1	-1				-1	-1		
6		1		-1				1	-1
7		-1	-1		-1		1		
8			1		-1	1			1
9			-1	1		1	-1		

R.4: $a(3) > 0$
R.3: $b(3) < 0$
R.2: $c(3) > 0$
R.6: $d(3) < 0$
R.1: $e(3) < 0$
contradiction in R.9.

6.19

12:345
36:478
14:78
17:29
25:68
34:59
58:79
6:13
9:26

	x	w	y	B	C	z	E	A	D
1	1		-1	1				-1	
2	1			-1	1				-1
3	-1	1				-1		-1	
4 *	-1	-1	-1			-1			
5	-1				1	1	-1		
6		1			-1			1	-1
7		-1	1	1			1		
8		-1	1		-1		-1		
9				-1		1	1		1

R.3: $a(1) < 0$
R.1: $b(1) < 0$
if $c(1) > 0$
R.2: $d(1) > 0$
contradiction in R.6;
hence $c(1) \leq 0$
R.8: $e(1) > 0$
contradiction in R.5.

6.20

12:345
36:478
14:78
25:67
27:89
39:56
48:59
1:26
9:13

	B	x	y	A	w	E	z	C	D
1	1		1					1	-1
2	1			1	1			-1	
3	-1	1				1			-1
4	-1	-1	1				-1		
5	-1			1		-1	1		
6		1		-1		-1		-1	
7		-1	-1	-1	1				
8 *		-1	-1		-1		-1		
9					-1	1	1		1

R.7: $a(1) < 0$
R.4: $b(1) < 0$
R.2: $c(1) < 0$
R.1: $d(1) < 0$
R.3: $e(1) < 0$
contradiction in R.6.

6.21

		B	x	A	y	z	D	w	C	E
12:345	1	1		1	1				-1	
36:478	2	1				1			1	-1
14:79	3	-1	1				1			-1
15:68	4	-1	-1	1				1		
29:78	5	-1			1		-1	-1		
39:56	6		1		-1		-1		-1	
47:58	7		-1	-1		-1		1		
2:16	8 *		-1		-1	-1		-1		
9:23	9			-1		1	1			1

R.7: $a(1) < 0$
R.4: $b(1) < 0$
R.1: $c(1) < 0$
R.6: $d(1) > 0$
R.2: $e(1) < 0$
contradiction in R.3.

6.22

		x	A	B	C	y	z	w	E	D
12:345	1	1		-1	1				-1	
36:478	2	1			-1	-1				-1
14:79	3	-1	1				1		-1	
17:28	4	-1	-1	-1				1		
25:68	5 *	-1				-1	-1	-1		
37:59	6		1			1			1	-1
49:58	7		-1	1	1		1			
6:13	8		-1		-1	1		-1		
9:26	9			1			-1	1		1

If $a(2) < 0$
R.8: $c(2) > 0$
R.7: $b(2) < 0$
R.3: $e(2) > 0$
contradiction in R.1;
hence $a(2) \geq 0$
R.4: $b(2) < 0$
R.3: $e(2) > 0$
R.6: $d(2) > 0$
R.2: $c(2) < 0$
contradiction in R.7.

6.23

		x	A	w	B	E	y	z	C	D
12:345	1	1		-1	1					-1
36:478	2	1			-1	1			-1	
15:67	3	-1	1				1		-1	
17:28	4	-1	-1			1		1		
24:89	5 *	-1		-1			-1	-1		
38:59	6		1	1					1	-1
49:57	7		-1	1	1			-1		
6:23	8		-1		-1	-1	1			
9:16	9					-1	-1	1		1

If $a(3) < 0$
R.3: $c(3) < 0$
R.6: $d(3) < 0$
R.1: $b(3) < 0$
R.2: $e(3) < 0$
contradiction in R.8;
hence $a(3) \geq 0$
R.7: $b(3) > 0$
R.1: $d(3) > 0$
R.3: $c(3) > 0$
R.2: $e(3) > 0$, contradiction in R.8.

6.24

		x	w	B	D	C	z	A	y	E
12:345										
36:478	1	1		1	1					-1
15:67	2	1			-1	1			-1	
17:28	3*	-1	-1				-1		-1	
25:89	4	-1	1				-1	1		
34:59	5	-1		1		1	1			
49:78	6		-1	-1					1	-1
6:23	7		1	-1	1			-1		
9:16	8		1		-1	-1		-1		
	9					-1	1	1		1

R.4: $a(1) > 0$
if $d(1) < 0$
R.7: $b(1) < 0$
R.2: $c(1) < 0$
contradiction in R.5;
hence $d(1) \geq 0$
R.8: $c(1) < 0$
R.5: $b(1) > 0$
R.7: $d(1) > 0$
R.1: $e(1) > 0$, contradiction in R.9.

6.25

		x	y	D	B	A	z	C	w	E
12:345										
36:478	1	1		1	1					-1
15:67	2	1			-1	-1			-1	
19:28	3*	-1	-1				-1		-1	
24:78	4	-1	1			-1	-1			
34:59	5	-1		1			1	1		
57:89	6		-1	-1					1	-1
6:23	7		1	-1		1		1		
9:16	8		1		-1	1		-1		
	9				1		1	-1		1

R.4: $a(1) < 0$
R.2: $b(1) > 0$
R.8: $c(1) < 0$
R.5: $d(1) > 0$
R.6: $e(1) < 0$
contradiction in R.1.

6.26

		x	y	B	C	D	z	w	A	E
12:345										
36:478	1	1		1	1				-1	
15:79	2	1			-1	1				-1
17:28	3	-1	1				-1		-1	
25:68	4*	-1	-1				-1	-1		
34:59	5	-1		1		1	1			
49:78	6		1			-1			1	-1
6:13	7		-1	-1	1			1		
9:26	8		-1		-1	-1		1		
	9			-1			1	-1		1

R.3: $a(1) < 0$
if $b(1) < 0$
R.5: $d(1) > 0$
R.8: $c(1) < 0$
R.6: $e(1) < 0$
contradiction in R.2;
hence $b(1) \geq 0$
R.1: $c(1) < 0$
R.7: $b(1) < 0$
R.9: $e(1) < 0$
R.2: $d(1) < 0$, contradiction in R.5.

Thus case #3 of Table 1 is impossible.

Chapter 7

In this chapter we treat the cases #6, 9 and #10. Here we examine first all the combinatorially different ways to embed $3K_{2,3} + 3K_{2,2}$ in K_9. We then check the graphs consisting of six edges of the type $K_9\backslash(3K_{2,3} + 3K_{2,2})$, to see if they decompose into three $K_{1,2}$ (for case #6), into $K_{2,2} + 2K_{1,1}$ (for case #9), or into $K_{2,2} + K_{1,2}$ (for case #10).

If two of the three $K_{2,3}$'s have equal pairs, then without loss of generality we may assume that $K_{2,3}(1) + K_{2,3}(2) = 12{:}345 + 12{:}678$. Thus $K_9\backslash(K_{2,3}(1) + K_{2,3}(2))$ is the amalgamation at the vertex 9 of a K_7 (having the vertices 3, 4, ..., 9) and a K_3 (having the vertices 1, 2 and 9). If $3K_{2,3} + 3K_{2,2}$ are to be embeded in K_9 in this situation, then the K_7 has to satisfy $K_7 = K_{2,3} + 3K_{2,2} + 3K_{1,1}$.

However, this is impossible, and we state it as a separate claim.

Claim: There exist no decompositions of K_7 of the form $K_{2,3} + 3K_{2,2} + 3K_{1,1}$.

Proof of the Claim: Suppose, on the contrary, that $K_7 = K_{2,3} + 3K_{2,2} + 3K_{1,1}$. Without loss of generality, let the vertices of K_7 be $\{1, 2, \ldots, 7\}$, and let $K_{2,3} = 12{:}345$. It follows from valence considerations that the three $K_{1,1}$'s must be either $1{:}6 + 2{:}7 + 6{:}7$, or else $1{:}7 + 2{:}6 + 6{:}7$, since K_7 and the $K_{2,2}$'s have only even valent vertices. In both cases, the edge $1{:}2$ is not a part of a $K_{2,2}$ in $K_7\backslash(K_{2,3} + 3K_{1,1})$. Therefore the said decomposition does not exist. This completes the proof of the Claim.

It follows that in these cases, no two of the $K_{2,3}$'s can have pairs which are equal.

Next, assume that two of the the three $K_{2,3}$'s have pairs which intersect in just one vertex. Thus, without loss of generality, we may assume that $K_{2,3}(1) = 12:345$ and that $K_{2,3}(2) = 13:678$ or 16:789. Both of these two cases lead, subject to the restrictions of Lemma 8 and 9, to no embeddings of $3K_{2,3} + 3K_{2,2}$ in K_9.

Therefore, it suffices to consider the case where the three $K_{2,3}$'s have pairs which are mutually disjoint. Using the notations of the previous chapter and Corollary 2, we may assume that $K_{2,3}(1) + K_{2,3}(2)$ are as in one of the following cases: 6*1, 6*2, 6*3, 6*4, 6*5, 6*11 and 6*12. In each one of these cases, we let the machine produce the list of all the possible $K_{2,3}(3)$. We marked all those $K_{2,3}(3)$ which are to be treated , making sure that those $K_{2,3}(3)$'s which are untreated will lead to isomorphic duplicates. In each one of the possibilities for $K_{2,3}(3)$, the machine lists all the possible $K_{2,2}$'s. Using again the idea of nested lists, we got all the possible $3K_{2,3} + 3K_{2,2}$ in K_9, subject to the established restrictions. The machine then drew the remaining graphs of the type $K_9 \setminus (3K_{2,3} + 3K_{2,2})$, which were found that way. As none of these graphs contains a $K_{2,2}$, it follows that cases #9 and #10 of Table 1 are impossible.

To check the situation for case #6, where the graphs of the form $K_9 \setminus (3K_{2,3} + 3K_{2,2})$ are supposed to be decomposable into three $K_{1,2}$'s, we add an extra check. This check if for the existance of an isolated edge. Only those cases having no isolated edges were listed, since three $K_{1,2}$'s have no isolated egdes. We then manually checked for isomorphic duplicates. The fifty cases obtained this way are presented, each one with its matrix and a coded proof, showing that it leads to a contradiction.

7.1

12:345
34:567
67:158
15:89
26:79
38:49
9:47
2:16
8:23

	B	x	y	C	w	E	z	A	D
1	1		1	1				-1	
2	1				1			1	-1
3	-1	1				1			-1
4	-1	1				-1	-1		
5	-1	-1	1	1					
6		-1	-1		1			-1	
7 *		-1	-1		-1		-1		
8			1	-1		1			1
9				-1	-1	-1	1		

R.6: $a(1) < 0$
if $c(1) < 0$
R.5: $b(1) < 0$
R.2: $d(1) < 0$
R.3: $e(1) < 0$
contradiction in R.4;
hence $c(1) \geq 0$
R.1: $b(1) < 0$
R.2: $d(1) < 0$
R.3: $e(1) < 0$
contradiction in R.4.

7.2

12:345
34:567
67:158
16:29
25:89
38:49
7:26
8:13
9:47

	x	y	w	A	z	D	B	C	
1	1		-1	1				-1	
2	1			-1	-1		-1		
3	-1	1				1		-1	
4	-1	1				-1			-1
5 *	-1	-1	-1		-1				
6		-1	1	1			-1		
7		-1	1				1		-1
8			-1		1	1		1	
9				-1	1	-1			1

If $a(1) < 0$
R.6: $b(1) < 0$
contradiction in R.2;
hence $a(1) \geq 0$
R.1: $c(1) > 0$
R.3: $d(1) > 0$
contradiction in R.8.

7.3

12:345
34:567
68:125
17:29
36:89
58:79
7:16
4:38
9:24

	x	C	y	w	B	E	A	D	z
1	-1		-1	1			-1		
2 *	-1		-1	-1					-1
3	1	1			1			-1	
4	1	1						1	-1
5	1	-1	-1			-1			
6		-1	1		1		-1		
7		-1		1		1	1		
8			1		-1	-1		-1	
9				-1	-1	1			1

R.1: $a(1) < 0$
If $b(1) > 0$
R.6: $c(1) > 0$
R.3: $d(1) > 0$
contradiction in R.4;
hence $b(1) \leq 0$
R.9: $e(1) < 0$
R.8: $d(1) > 0$
R.4: $c(1) < 0$
contradiction in R.5.

7.4

12:345
34:567
68:125
17:29
38:49
56:79
9:24
8:36
7:18

	x	y	B	C	w	D	A	z	
1	-1		-1	1					-1
2	-1		-1	-1			-1		
3 *	1	1			1			1	
4	1	1			-1		-1		
5	1	-1	-1			1			
6		-1	1			1		1	
7		-1		1		-1			1
8			1		1			-1	-1
9				-1	-1	-1	1		

R.4: $a(1) > 0$
If $b(1) > 0$
R.2: $c(1) < 0$
R.6: $d(1) < 0$
contradiction in R.9;
hence $b(1) \leq 0$
R.5: $d(1) < 0$
contradiction in R.6.

7.5

12:345
34:567
68:129
17:29
35:89
58:67
7:16
9:24
4:38

	B	x	y	A	E	z	w	C	D
1	1		1	1			-1		
2	1		1	-1				-1	
3	-1	1			1				-1
4	-1	1						-1	1
5	-1	-1			1	1			
6 *		-1	-1			-1	-1		
7		-1		1		-1	1		
8			-1		-1	1			-1
9			1	-1	-1			1	

R.7: $a(1) > 0$
R.1: $b(1) < 0$
R.2: $c(1) < 0$
R.4: $d(1) < 0$
R.3: $e(1) < 0$
contradiction in R.8.

7.6

12:345
34:567
68:157
15:79
29:67
38:49
2:18
8:36
9:24

	B	x	w	y	z	E	C	A	D
1	1		-1	1			-1		
2	1				1		1		-1
3	-1	1				1		-1	
4	-1	1				-1			-1
5	-1	-1	-1	1					
6		-1	1		-1			-1	
7 *		-1	-1	-1	-1				
8			1			1	-1	1	
9				-1	1	-1			1

R.6: $a(2) < 0$
R.5: $b(2) > 0$
R.1: $c(2) > 0$
R.2: $d(2) > 0$
R.4: $e(2) < 0$
contradiction in R.8.

7.7

12:345
34:567
68:157
15:79
29:78
36:89
2:16
4:38
9:24

	A	x	y	w	z	D	B	E	C
1	1		-1	1			-1		
2	1				1		1		-1
3	-1	1				1		-1	
4	-1	1						1	-1
5	-1	-1	-1	1					
6		-1	1			1	-1		
7 *		-1	-1	-1	-1				
8			1		-1	-1		-1	
9				-1	1	-1			1

R.5: $a(2) < 0$
R.1: $b(2) < 0$
R.2: $c(2) < 0$
R.6: $d(2) < 0$
R.4: $e(2) < 0$
contradiction in R.8.

7.8

12:345
34:567
68:159
15:79
26:78
37:89
2:16
4:38
9:24

	B	x	E	y	z	w	D	A	C
1	1		-1	1			-1		
2	1				1		1		-1
3	-1	1				-1		-1	
4	-1	1						1	-1
5	-1	-1	-1	1					
6		-1	1		1		-1		
7 *		-1		-1	-1	-1			
8			1		-1	1		-1	
9			-1	-1		1			1

If $a(3) \le 0$
R.3: $b(3) > 0$
R.4: $c(3) < 0$
R.2: $d(3) < 0$
R.5: $e(3) < 0$
contradiction in R.1;
hence $a(3) > 0$
R.8: $e(3) > 0$
R.6: $d(3) > 0$
R.1: $b(3) > 0$
contradiction in R.5.

7.9

12:345
34:567
68:159
16:27
25:79
37:89
4:38
8:26
9:14

	x	y	B	C	E	A	w	D	z
1	1		-1	-1					-1
2	1			1	-1			-1	
3	-1	-1				1	1		
4 *	-1	-1					-1		-1
5	-1	1	-1		-1				
6		1	1	-1				-1	
7		1		1	1	1			
8			1			-1	1	1	
9			-1		1	-1			1

R.3: $a(2) > 0$
if $b(2) > 0$
R.1: $c(2) < 0$
R.6: $d(2) > 0$
R.2: $e(2) < 0$
contradiction in R.9;
hence $b(2) \le 0$
R.1: $d(2) > 0$
R.8: $c(2) > 0$
R.5: $e(2) > 0$
contradiction in R.7.

7.10

		x	y	C	E	B	A	w	D	z
12:345	1	1		-1	1					-1
34:567	2	1			-1	1			-1	
68:179	3	-1	-1				1	1		
19:27	4 *	-1	-1					-1		-1
25:67	5	-1	1			1	1			
35:89	6		1	1		-1			-1	
4:38	7		1	-1	-1	-1				
8:26	8			1			-1	1	1	
9:14	9			-1	1		-1			1

R.3: $a(2) > 0$
R.5: $b(2) < 0$
if $c(2) > 0$
R.6: $d(2) > 0$
R.1: $e(2) > 0$
contradiction in R.2;
hence $c(2) \le 0$
R.7: $e(2) > 0$
contradiction in R.1.

7.11

		B	x	E	w	y	z	D	A	C
12:345	1	1		-1	1				-1	
34:567	2	1			-1	1				-1
89:135	3	-1	1	-1				-1		
16:27	4	-1	1					1		-1
28:79	5	-1	-1	-1			1			
59:67	6		-1		1		-1		1	
4:39	7 *		-1		-1	-1	-1			
6:18	8			1		1			-1	1
8:24	9			1		-1	1	-1		

R.6: $a(2) > 0$
if $b(2) > 0$
R.2: $c(2) > 0$
R.4: $d(2) > 0$
R.3: $e(2) < 0$
contradiction in R.9;
hence $b(2) \le 0$
R.1: $e(2) < 0$
contradiction in R.5.

7.12

		B	x	C	A	y	w	E	z	D
12:345	1	1		-1	1				1	
34:567	2	1			-1	1				-1
89:135	3	-1	1	-1				-1		
18:26	4	-1	1					1		-1
25:67	5	-1	-1	-1		1				
68:79	6		-1		-1	-1	1			
4:38	7 *		-1			-1	-1		-1	
7:19	8			1	1		1	-1		
9:24	9			1			-1		1	1

R.6: $a(2) < 0$
if $b(2) < 0$
R.1: $c(2) < 0$
contradiction in R.5;
hence $b(2) \ge 0$
R.2: $d(2) > 0$
R.4: $e(2) > 0$
R.8: $c(2) > 0$
contradiction in R.9.

7.13

		x	y	z	D	B	C	w		A
12:345	1	1		-1	1				-1	
34:567	2	1			-1	1				-1
89:136	3 *	-1	-1	-1				-1		
17:26	4	-1	-1					1		-1
25:68	5	-1	1			1	1			
58:79	6		1	-1	-1	-1				
4:38	7		1		1		-1		1	
7:19	8			1		-1	1	-1		
9:24	9			1			-1		-1	1

R.4: a(3) < 0
if b(3) < 0
R.5: c(3) > 0
contradiction in R.8;
hence b(3) ≥ 0
R.5: c(3) < 0
R.2: d(3) > 0
contradiction in R.6.

7.14

		A	x	y	w	D	z	B		C
12:345	1	1			1	1				-1
34:567	2	1				-1	1		-1	
89:356	3	-1	1	-1				-1		
15:67	4	-1	1					1		-1
17:28	5	-1	-1	-1	1					
27:69	6 *		-1	-1	-1		-1			
4:38	7		-1		-1	1	1			
8:29	8			1		-1		-1	1	
9:14	9			1			-1		-1	1

R.5: a(3) < 0
R.3: b(3) > 0
R.4: c(3) > 0
R.1: d(3) > 0
contradiction in R.7.

7.15

		B	x	y	z	A	w	D	C	
12:345	1	1			1	1				-1
34:567	2	1				-1	1		-1	
89:356	3	-1	1	-1				-1		
15:67	4	-1	1					1	-1	
17:28	5	-1	-1	-1	1					
27:69	6 *		-1	-1	-1		-1			
4:39	7		-1		-1	1	1			
8:24	8			1		-1			1	-1
9:18	9			1			-1	-1		1

R.7: a(1) > 0
R.5: b(1) < 0
R.2: c(1) < 0
R.3: d(1) > 0
contradiction in R.4.

7.16

12:345
34:567
89:356
15:67
18:29
27:69
4:39
7:28
8:14

	A	x	y	z	C	w	E	B	D
1	1			1	1				-1
2	1				-1	1		-1	
3	-1	1	-1				-1		
4	-1	1					1		-1
5	-1	-1	-1	1					
6 *		-1	-1	-1		-1			
7		-1		-1		1		1	
8			1		1			-1	1
9			1		-1	-1	-1		

R.5: $a(1) < 0$
R.7: $b(1) > 0$
R.2: $c(1) < 0$
R.1: $d(1) < 0$
R.3: $e(1) > 0$
contradiction in R.4.

7.17

12:345
34:567
89:367
15:68
16:27
25:79
4:38
8:29
9:14

	B	x	y	A	w	z	D		C
1	1			1	1				-1
2	1				-1	1		-1	
3	-1	1	-1				-1		
4	-1	1					1		-1
5	-1	-1		1		1			
6		-1	-1	-1	1				
7 *		-1	-1		-1	-1			
8			1	-1			-1	1	
9			1			-1		-1	1

R.6: $a(1) < 0$
R.5: $b(1) < 0$
R.1: $c(1) < 0$
R.4: $d(1) < 0$
contradiction in R.3.

7.18

12:345
34:567
89:367
15:68
16:27
25:79
4:39
8:24
9:18

	B	x	y	A	w	z	E	D	C
1	1			1	1				-1
2	1				-1	1		-1	
3	-1	1	-1				-1		
4	-1	1					1	-1	
5	-1	-1		1		1			
6		-1	-1	-1	1				
7 *		-1	-1		-1	-1			
8			1	-1				1	-1
9			1			-1	-1		1

R.6: $a(1) < 0$
R.5: $b(1) < 0$
R.1: $c(1) < 0$
R.8: $d(1) < 0$
R.3: $e(1) > 0$
contradiction in R.4.

7.19

		B	x	y	A	w	z	D	C	
12:345	1	1			1	1			-1	
34:567	2	1				-1	1			-1
89:367	3	-1	1	-1				-1		
15:69	4	-1	1					1	-1	
16:27	5	-1	-1		1		1			
25:78	6		-1	-1	-1	1				
4:39	7 *		-1	-1		-1	-1			
8:14	8			1			-1		1	-1
9:28	9			1	-1			-1		1

R.6: a(1) < 0
R.5: b(1) < 0
R.1: c(1) < 0
R.3: d(1) > 0
contradiction in R.4.

7.20

		x	C	y	w	B	E	D	A	z
12:345	1	-1		-1	1				-1	
34:678	2 *	-1		-1	-1					-1
67:129	3	1	1			1		-1		
18:29	4	1	1			-1				-1
35:49	5	1				1	1	1		
56:78	6		-1	1			1	-1		
5:36	7		-1	1			-1		-1	
8:17	8		-1		1		-1		1	
9:24	9			-1	-1	-1				1

R.1: a(3) < 0
R.9: b(3) > 0
R.4: c(3) > 0
R.3: d(3) > 0
R.5: e(3) < 0
contradiction in R.6.

7.21

		B	x	y	C	z	w	D	E	A
12:345	1	1			1	1		-1		
36:457	2	1					1	1		-1
58:479	3	-1	1		1					-1
13:68	4	-1	-1	-1		1				
14:79	5	-1	-1	1					1	
29:67	6		1		-1		-1		1	
2:18	7 *		-1	-1		-1	-1			
8:56	8			1	-1			-1	-1	
9:23	9			-1		-1	1			1

R.9: a(3) > 0
R.4: b(3) > 0
R.3: c(3) > 0
R.1: d(3) > 0
R.5: e(3) > 0
contradiction in R.8.

7.22

		B	x	y	A	w	z	C	D	E
12:345										
36:457	1	1			1	1		-1		
58:479	2	1					1	1	-1	
13:89	3	-1	1		1				-1	
19:67	4	-1	-1	-1			1			
24:79	5	-1	-1	1						-1
2:18	6		1			-1			1	-1
6:23	7 *		-1	-1		-1	-1			
8:56	8			1	-1			-1		1
	9			-1	-1	1	-1			

R.9: $a(2) < 0$
R.4: $b(2) < 0$
R.1: $c(2) < 0$
R.2: $d(2) < 0$
R.5: $e(2) > 0$
contradiction in R.8.

7.23

		B	x	y	A	w	z	C		D
12:345										
36:457	1	1			1	1		-1		
58:479	2	1					1	1		-1
13:89	3	-1	1		1				-1	
19:67	4	-1	-1	-1			1			
24:79	5	-1	-1	1						-1
2:16	6		1			-1		-1	1	
6:38	7 *		-1	-1		-1	-1			
8:25	8			1	-1				-1	1
	9			-1	-1	1	-1			

R.9: $a(2) < 0$
R.4: $b(2) < 0$
R.1: $c(2) < 0$
R.2: $d(2) < 0$
contradiction in R.5.

7.24

		B	x	y	z	C	w		D	A
12:345										
36:457	1	1			1	1			-1	
58:479	2	1				-1	1			-1
14:79	3	-1	1					1		-1
18:26	4	-1	-1	-1	1					
29:67	5	-1	-1	1					-1	
3:68	6		1			-1	-1	-1		
8:15	7 *		-1	-1	-1		-1			
9:23	8			1		1		-1	1	
	9			-1	-1		1			1

R.9: $a(2) > 0$
R.4: $b(2) < 0$
R.2: $c(2) < 0$
R.1: $d(2) < 0$
contradiction in R.5.

7.25

12:345
36:457
78:124
13:69
47:59
56:89
2:16
8:37
9:28

	x	y	A	E	w	z	D	B	C
1	1		-1	1			-1		
2	1		-1				1		-1
3	-1	1		1				-1	
4	-1	-1	-1		1				
5 *	-1	-1			-1	-1			
6		1		-1		-1	-1		
7		-1	1		1			-1	
8			1			1		1	-1
9				-1	-1	1			1

R.4: a(2) < 0
R.7: b(2) < 0
R.8: c(2) < 0
R.2: d(2) < 0
R.1: e(2) < 0
contradiction in R.6.

7.26

12:345
36:457
78:124
16:29
35:89
47:59
6:13
8:67
9:28

	x	y	z	C	A	w	B		D
1	1		-1	1			-1		
2	1		-1	-1					-1
3	-1	1			1		-1		
4 *	-1	-1	-1			-1			
5	-1	-1			1	1			
6		1		1			1	-1	
7		-1	1			-1		-1	
8			1		-1			1	-1
9				-1	-1	1			1

R.5: a(2) > 0
R.3: b(2) > 0
R.1: c(2) > 0
R.2: d(2) < 0
contradiction in R.9.

7.27

12:345
36:457
89:134
18:26
25:79
48:57
6:23
7:19
9:68

	x	y	z	C	A	w	D	B	E
1	1		-1	1				-1	
2	1			-1	1		-1		
3	-1	1	-1				-1		
4 *	-1	-1	-1			-1			
5	-1	-1			1	1			
6		1		-1			1		-1
7		-1			-1	1		1	
8			1	1		-1			-1
9			1		-1			-1	1

R.5: a(3) > 0
R.7: b(3) > 0
R.1: c(3) > 0
R.3: d(3) < 0
R.6: e(3) < 0
contradiction in R.8.

7.28

		A	x	B	y	w	z	C	D	E
12:345	1	1		-1	1			-1		
36:457	2	1				1		1	-1	
89:134	3	-1	1	-1					-1	
18:67	4	-1	-1	-1			1			
25:78	5	-1	-1			1	-1			
49:57	6		1		-1				1	-1
2:19	7*		-1		-1	-1	1			
6:23	8			1	1	-1				-1
9:68	9			1			1	-1		1

R.5: $a(3) < 0$
R.4: $b(3) > 0$
R.1: $c(3) < 0$
R.2: $d(3) < 0$
R.6: $e(3) < 0$
contradiction in R.8.

7.29

		x	y	z	C	A	w	D		B
12:345	1	1		-1	1			-1		
36:457	2	1				1		1	-1	
89:134	3	-1	1	-1						-1
18:67	4*	-1	-1	-1			-1			
25:78	5	-1	-1			1	1			
49:57	6		1		-1			-1		1
2:16	7		-1		-1	-1	1			
9:28	8			1	1	-1			-1	
6:39	9			1			-1		1	-1

R.5: $a(2) > 0$
R.3: $b(2) > 0$
R.7: $c(2) < 0$
R.1: $d(2) < 0$
contradiction in R.6.

7.30

		x	y	z	A	E	w	D	B	C
12:345	1	1		-1				1		-1
36:457	2	1			1	1		-1		
89:134	3	-1	1	-1					-1	
25:78	4*	-1	-1	-1			-1			
28:69	5	-1	-1		1		1			
49:57	6		1			-1		-1	1	
1:26	7		-1		-1		1			1
6:39	8			1	-1	1				-1
7:18	9			1		-1	-1		-1	

R.5: $a(3) > 0$
R.3: $b(3) < 0$
R.7: $c(3) > 0$
R.1: $d(3) > 0$
R.6: $e(3) < 0$
contradiction in R.9.

7.31

12:345
36:457
89:146
23:68
25:79
48:57
1:26
7:19
9:38

	x	y	A	D	w	z	C	B	E
1	1		-1				1	-1	
2	1			1	-1		-1		
3	-1	1		1					-1
4	-1	-1	-1			1			
5 *	-1	-1			-1	-1			
6		1	-1	-1			-1		
7		-1			1	-1		1	
8			1	-1		1			-1
9			1		1			-1	1

R.4: $a(3) > 0$
R.7: $b(3) > 0$
R.1: $c(3) > 0$
R.6: $d(3) < 0$
R.3: $e(3) < 0$
contradiction in R.8.

7.32

12:345
36:457
89:346
15:78
18:29
49:57
2:69
6:13
7:28

	x	y	z	A	C	w	D	B	
1	1			1	1			-1	
2	1				-1		1		-1
3	-1	1	-1					-1	
4 *	-1	-1	-1			-1			
5	-1	-1		1		1			
6		1	-1				-1	1	
7		-1		-1		1			1
8			1	-1	1				-1
9			1		-1	-1	-1		

R.5: $a(3) > 0$
R.3: $b(3) < 0$
R.1: $c(3) < 0$
R.6: $d(3) < 0$
contradiction in R.9.

7.33

12:345
36:457
89:346
15:78
28:79
49:57
1:69
2:18
6:23

	x	y	z	A	D	w	C		B
1	1			1			1	-1	
2	1				1			1	-1
3	-1	1	-1						-1
4 *	-1	-1	-1			-1			
5	-1	-1		1		1			
6		1	-1				-1		1
7		-1		-1	-1	1			
8			1	-1	1			-1	
9			1		-1	-1	-1		

R.5: $a(3) > 0$
R.3: $b(3) < 0$
R.6: $c(3) < 0$
R.7: $d(3) < 0$
contradiction in R.9.

7.34

12:345
36:457
89:367
14:78
25:79
48:59
1:29
2:68
6:13

	x	y	B	A	z	w		C	D
1	1			1			1		-1
2	1				-1		-1	1	
3	-1	1	-1						-1
4	-1	-1		1		1			
5*	-1	-1			-1	-1			
6		1	-1					-1	1
7		-1	-1	-1	1				
8			1	-1		1		-1	
9			1		1	-1	-1		

R.4: a(2) > 0
R.7: b(2) < 0
R.8: c(2) < 0
R.6: d(2) < 0
contradiction in R.3.

7.35

12:345
36:457
89:367
14:78
25:79
48:59
1:29
2:68
6:13

	x	y	B	A	z	w		C	D
1	1			1			1		-1
2	1				-1		-1	1	
3	-1	1	-1						-1
4	-1	-1		1		1			
5*	-1	-1			-1	-1			
6		1	-1					-1	1
7		-1	-1	-1	1				
8			1	-1		1		-1	
9			1		1	-1	-1		

R.4: a(2) > 0
R.7: b(2) < 0
R.8: c(2) < 0
R.6: d(2) < 0
contradiction in R.3.

7.36

12:345
36:457
89:367
14:78
25:79
48:59
1:69
2:18
6:23

	x	y	B	A	z	w	D	C	E
1	1			1			1	-1	
2	1				-1			1	-1
3	-1	1	-1						-1
4	-1	-1		1		1			
5*	-1	-1			-1	-1			
6		1	-1				-1		1
7		-1	-1	-1	1				
8			1	-1		1		-1	
9			1		1	-1	-1		

R.4: a(2) > 0
R.7: b(2) < 0
R.8: c(2) < 0
R.1: d(2) < 0
R.2: e(2) < 0
contradiction in R.3.

7.37

		A	x	y	w	D	z	E	C	B
12:345	1	1			1	1		-1		
36:478	2	1					1	1		-1
57:489	3	-1	1						1	-1
14:89	4	-1	-1	-1	1					
15:67	5	-1		1		1			-1	
29:68	6		1			-1	-1		-1	
2:17	7		-1	1		-1		-1		
3:56	8 *		-1	-1	-1		-1			
9:23	9			-1	-1		1			1

R.4: a(3) < 0
R.9: b(3) < 0
R.3: c(3) < 0
R.5: d(3) < 0
R.7: e(3) > 0
contradiction in R.1.

7.38

		x	w	y	A	B	C		z	D
12:345	1	1			-1	-1			-1	
36:478	2	1				1		1		-1
59:346	3 *	-1	-1	-1					-1	
14:78	4	-1	1	-1	-1					
17:29	5	-1		1			1			-1
58:79	6		-1	-1				-1	1	
2:69	7		1		1	-1	-1			
6:13	8		1		1		1			1
8:25	9			1		1	-1	-1		

R.4: a(1) < 0
R.1: b(1) > 0
R.7: c(1) < 0
R.8: d(1) > 0
contradiction in R.5.

7.39

		x	w	y	C	E	z	B	A	D
12:345	1	1			1	1			-1	
36:478	2	1				-1		1		-1
59:346	3	-1	1	-1					-1	
15:79	4 *	-1	-1	-1			-1			
17:28	5	-1		1	1					-1
49:78	6		1	-1				-1	1	
2:69	7		-1		-1	1	1			
6:13	8		-1			-1	1			1
8:25	9			1	-1		-1	-1		

R.3: a(2) < 0
R.6: b(2) < 0
R.9: c(2) > 0
R.5: d(2) > 0
R.2: e(2) < 0
contradiction in R.8.

7.40

		A	x	y	w	D	z	E	B	C
12:345	1	1			1	1		-1		
36:478	2	1					1	1		-1
59:347	3	-1	1	-1					-1	
14:78	4	-1	-1	-1	1					
15:69	5	-1		1		1				-1
28:79	6		1			-1		-1	1	
2:16	7 *		-1	-1	-1		-1			
6:39	8		-1		-1		1			1
8:25	9			1		-1	-1		-1	

R.4: $a(3) < 0$
R.3: $b(3) > 0$
R.8: $c(3) < 0$
R.9: $d(3) < 0$
R.1: $e(3) < 0$
contradiction in R.6.

7.41

		x	w	y	z	B	E	C	A	D
12:345	1	1			-1	1			-1	
36:478	2	1				-1	1			-1
59:347	3	-1	1	-1					-1	
14:78	4 *	-1	-1	-1	-1					
16:29	5	-1		1				1		-1
28:79	6		1			1		-1	1	
5:69	7		-1	-1	1		-1			
6:13	8		-1		1		1			1
8:25	9			1		-1	-1	-1		

R.3: $a(1) < 0$
R.1: $b(1) < 0$
R.6: $c(1) < 0$
R.5: $d(1) < 0$
R.2: $e(1) < 0$
contradiction in R.7.

7.42

		x	w	y	z	C	B	D	A	
12:345	1	1			-1			1		-1
36:478	2	1				1	1	-1		
59:347	3	-1	1	-1					-1	
14:78	4 *	-1	-1	-1	-1					
25:68	5	-1		1		1				-1
28:79	6		1			-1		-1	1	
1:26	7		-1	-1	1		-1			
6:39	8		-1		1	-1	1			
9:15	9			1			-1		-1	1

R.3: $a(1) < 0$
R.7: $b(1) > 0$
R.8: $c(1) > 0$
R.6: $d(1) < 0$
contradiction in R.2.

7.43

12:345
36:478
59:347
14:78
25:68
28:79
1:29
6:13
9:56

	x	w	y	z	D	C	B	A	
1	1			-1			1	-1	
2	1				1	1	-1		
3	-1	1	-1					-1	
4 *	-1	-1	-1	-1					
5	-1		1		1				-1
6		1			-1			1	-1
7		-1	-1	1		-1			
8		-1		1	-1	1			
9			1			-1	-1		1

R.3: $a(1) < 0$
R.1: $b(1) < 0$
R.7: $c(1) > 0$
R.2: $d(1) < 0$
contradiction in R.8.

7.44

12:345
36:478
59:367
14:79
17:28
49:58
2:69
6:13
8:25

	A	w	x	y	z	E	C	B	D
1	1			1	-1			-1	
2	1				1		1		-1
3	-1	1	-1					-1	
4	-1	-1		1		1			
5	-1		1			-1			-1
6		1	-1				-1	1	
7 *		-1	-1	-1	-1				
8		-1			1	-1			1
9			1	-1		1	-1		

If $a(1) > 0$
R.3: $b(1) < 0$
contradiction in R.1;
hence $a(1) \leq 0$
R.1: $b(1) < 0$
R.6: $c(1) < 0$
R.2: $d(1) < 0$
R.4: $e(1) < 0$
contradiction in R.5.

7.45

12:345
36:478
59:467
13:69
14:78
28:79
2:16
5:39
8:25

	B	x	y	D	z	w	E	C	A
1	1			1	1		-1		
2	1					1	1		-1
3	-1	1		1				-1	
4	-1	-1	-1		1				
5	-1		1					1	-1
6		1	-1	-1			-1		
7 *		-1	-1		-1	-1			
8		-1			-1	1			1
9			1	-1		-1		-1	

R.8: $a(3) > 0$
R.4: $b(3) > 0$
R.5: $c(3) > 0$
R.3: $d(3) > 0$
R.1: $e(3) > 0$
contradiction in R.6.

7.46

		x	C	y	z	D	B	w	E	A
12:345	1 *	1		1	1			1		
36:478	2	1		1				-1		-1
79:124	3	-1	1			1			-1	
15:68	4	-1	-1	1			1			
37:59	5	-1			1	-1	-1			
49:58	6		1		-1			1	1	
2:16	7		-1	-1		1				-1
6:39	8		-1		-1		-1			1
8:27	9			-1		-1	1		-1	

R.2: $a(1) > 0$
if $b(1) < 0$
R.4: $c(1) < 0$
contradiction in R.8;
hence $b(1) \geq 0$
R.5: $d(1) < 0$
R.7: $c(1) < 0$
R.3: $e(1) < 0$
contradiction in R.6.

7.47

		x	D	y	A	B	C	w	z	E
12:345	1	1		1	1			-1		
36:478	2 *	1		1				1	1	
79:124	3	-1	1			1				-1
15:68	4	-1	-1	1			1			
39:56	5	-1			1	-1	-1			
47:58	6		1		-1	-1		-1		
2:16	7		-1	-1			1			-1
8:29	8		-1		-1		-1		-1	
9:37	9			-1		1			1	1

R.1: $a(1) < 0$
if $b(1) > 0$
R.5: $c(1) < 0$
R.8: $d(1) > 0$
contradiction in R.4;
hence $b(1) \leq 0$
R.9: $e(1) > 0$
R.3: $d(1) > 0$
R.4: $c(1) > 0$
contradiction in R.6.

7.48

		x	D	y	A	B	C	w	E	z
12:345	1	1		1	1			-1		
36:478	2 *	1		1				1		1
79:125	3	-1	1			1			-1	
15:68	4	-1	-1			1	1			
34:59	5	-1		1	1	-1				
49:78	6		1		-1			-1	1	
2:16	7		-1	-1			-1			1
6:39	8		-1		-1		-1			-1
8:27	9			-1		-1	1		-1	

R.1: $a(3) < 0$
R.5: $b(3) < 0$
if $c(3) < 0$
R.4: $d(3) < 0$
R.3: $e(3) < 0$
contradiction in R.7;
hence $c(3) \geq 0$
R.9: $e(3) > 0$
R.6: $d(3) < 0$
contradiction in R.3.

7.49

		x	w	D	C	y	z	B	A	E
12:345	1	1		-1	1			-1		
36:478	2	1				-1		1	-1	
79:158	3	-1	1				-1		-1	
15:68	4 *	-1	-1			-1	-1			
24:78	5	-1		-1	1		1			
34:59	6		1		-1				1	-1
2:19	7		-1	1		1				-1
6:23	8		-1	-1	-1	1				
9:67	9			1			1	-1		1

R.3: $a(3) < 0$
R.2: $b(3) < 0$
if $c(3) > 0$
R.5: $d(3) > 0$
R.6: $e(3) < 0$
contradiction in R.7;
hence $c(3) \leq 0$
R.8: $d(3) > 0$
R.7: $e(3) > 0$
contradiction in R.9.

7.50

		x	w	C	B	y	z	D	A	E
12:345	1	1		-1	1			-1		
36:478	2	1				-1		1		-1
79:158	3	-1	1				-1		-1	
15:68	4 *	-1	-1			-1	-1			
24:78	5	-1		-1	1		1			
34:59	6		1		-1			-1	1	
2:16	7		-1	1		1				-1
6:39	8		-1	-1	-1	1				
9:27	9			1			1		-1	1

R.3: $a(2) < 0$
if $b(2) > 0$
R.5: $c(2) > 0$
R.6: $d(2) < 0$
R.2: $e(2) < 0$
contradiction in R.7;
hence $b(2) \leq 0$
R.5: $c(2) < 0$
contradiction in R.8.

Thus cases #6, #9 and #10 are impossible.

Chapter 8

In this chapter we treat the cases #12, 16, 17 and #18.

To ease our search by computer, we treat all of these cases together, by trying first to embed $4K_{2,3} + K_{2,2}$ in K_9, subject to the established restrictions.

In the first part, we assume that the four $K_{2,3}$'s have two pairs having a non-empty intersection. Thus we may assume, without loss of generality, that $K_{2,3}(1) = 12{:}345$ and $K_{2,3}(2)$ is one of 12:678, 13:678 and 16:789.

If $K_{2,3}(1) + K_{2,3}(2) = 12{:}345 + 12{:}678$, then $K_9 \backslash (12{:}345 + 12{:}678)$ is the amalgamation of a K_3 and a K_7 at one vertex.

In case #12, one of the four $K_{1,2}$'s must be 1:29 or 2:19, implying that there exists a vertex of type (3, 3, 2, 0), but in this case there exist no such vertices, since $x_{0,1} = \alpha = 0$ (see equation (5)). Considering cases #16, 17 and #18, it follows that K_7 (on the vertices 3, 4, ..., 9) must have a decomposition into $2K_{2,3} + 2K_{2,2} + K_{1,1}$, contrary to the Graham-Pollak Theorem [7]. Thus $K_{2,3}(2) \neq 12{:}678$.

In the remaining cases, where $K_{2,3}(2) = 13{:}678$ or $16{:}789$, we let the computer present graphically all the possible graphs of the form $K_9 \backslash (4K_{2,3} + K_{2,2})$, subject to the established restrictions. If such a graph on eight edges has an isolated edge, then it is not related to case #12, since it is supposed to be decomposed into four $K_{1,2}$'s. If it has no $K_{2,2}$'s, then it is not related to cases #16, 17 or #18. Thus we disregard all those possibilities which contain an isolated edge and contain no $K_{2,2}$'s. We checked (manually) all the cases obtained this way, and dropped isomorphic duplicates. We checked manually in each one of the cases, to see if the graph on eight edges can be decomposed into either four $K_{1,2}$'s (for case #12), into $K_{2,2} + 4K_{1,1}$ (for case #16), $K_{2,2} + K_{1,2} + 2K_{1,1}$ (for case #17) or $K_{2,2} + 2K_{1,2}$ (for case #18). Whenever we had to consider a decomposition, related to case #12, it turned out that there exists a vertex of type (3, 3, 2, 0), contrary to property (5), because $x_{0,1} = \alpha = 0$. In most of the cases where a decomposition was related to cases #16, 17 or #18, we found two vertices u and v, such that m(u,v) = 3, contrary to Lemma 9. These cases were, therefore, deleted. In only two cases we had to apply the technique of treating the matrix, as follows.

8.1

12:345
13:678
45:369
78:269
16:29

and the remaining graph is:

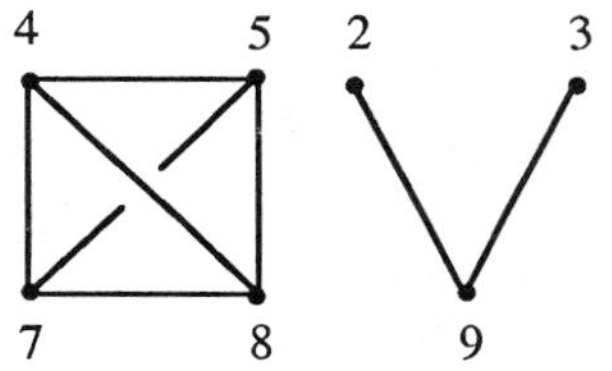

4 5 2 3
7 8 9

Vertex 1 is of type (3, 3, 2, 0), hence it suffices to look for a decomposition of the types as in cases #17 or #18. However, no such a decomposition exists for case #18, because the graph K_4 cannot be decomposed into $K_{2,2} + K_{1,2}$. For case #17, if $K_{2,2}(2)$ = 45:78, then m(4,5) = 3 (=m(7,8)), contrary to Lemma 9. If $K_{2,2}(2)$ = 47:58, then we get the following matrix, using $K_{1,2}$ = 9:23 and $2K_{1,1}$ = 4:7+5:8 (the column, corresponding to the plane containing the free facet of P_1 does not appear).

	B	A	x	y	w	C	z	D	E
1	1	1			1				
2	1			-1	-1		1		
3	-1	1	-1				1		
4	-1		1			1		1	
5	-1		1			-1			1
6		-1	-1	-1	1				
7		-1		1		1		-1	
8		-1		1		-1			-1
9 *			-1	-1	-1		-1		

R.6: $a(1) < 0$
R.2: $b(1) < 0$
if $c(1) > 0$
R.4: $d(1) < 0$
contradiction in R.7;
hence $c(1) \leq 0$
R.8: $e(1) > 0$
contradiction in R.5.

If $K_{2,2}(2) = 48:57$, then $2K_{1,1} = 4:8 + 5:7$, and we get an isomorphic duplicate of the previous case, obtained by the permutation (78); it leads to a similar contradiction.

8.2

12:345
16:789
34:567
89:257
25:67

and the remaining graph is:

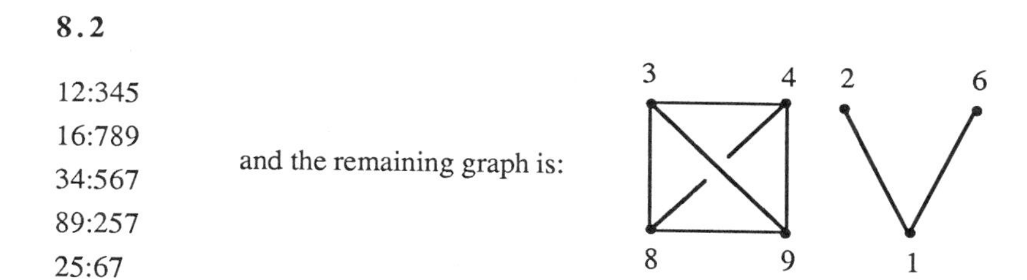

For case #12, one of the four $K_{1,2}$ must be 1:26, implying that the vertex 1 is of type (3, 3, 2, 0), contrary to property (5), because $x_{0,1} = \alpha = 0$. Case #18 is irrelevant, since K_4 is not decomposable into $K_{2,2} + K_{1,2}$.

For cases #16 or #17: if $K_{2,2} = 34:89$, then $m(3,4) = 3$ $(= m(8,9))$, contrary to Lemma 9. If $K_{2,2} = 38:49$, then $2K_{1,1} = 3:8 + 4:9$, leading to the following (partial) matrix.

	A	x	y	z	w	B	C	D
1	1	1						
2	1			-1	1			
3	-1		1			1	1	
4	-1		1			-1		1
5	-1		-1	-1	1			
6		1	-1		-1			
7 *		-1	-1	-1	-1			
8		-1		1		1	-1	
9		-1		1		-1		-1

R.5: $a(2) < 0$
if $b(2) > 0$
R.3: $c(2) < 0$
contradiction in R.8;
hence $b(2) \leq 0$
R.4: $d(2) < 0$
contradiction in R.9.

If $K_{2,2}$ = 39:48, then the two $K_{1,1}$ are 3:9 and 4:8, and the decomposition is isomorphic, by the permutation (89), to the previous one; it leads to a similar contradiction.

8.3

12:345
16:789
34:567
89:257
38:49

and the remaining graph is:

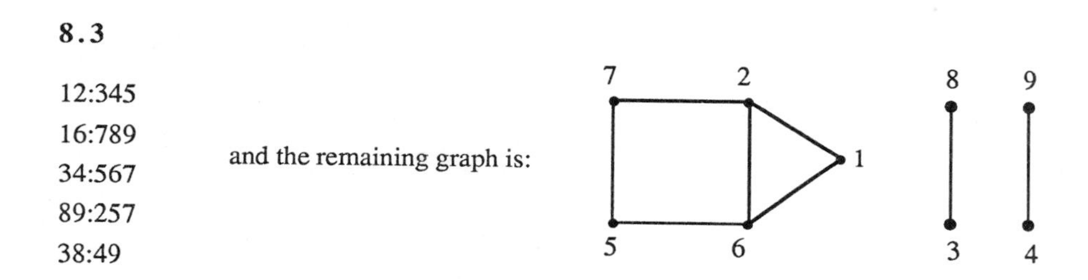

Cases #12 and #18 are irrelevant, because of the isolated edges.

For cases #16 and #17, the other $K_{2,2}$ can only be 25:67. Thus the decomposition contains, besides the four $K_{2,3}$'s (as in case 8.2), the following $2K_{2,2} + 2K_{1,1}$ = 25:67 + 38:49 + 3:8 + 4:9 (as in case 8.2), which implies a contradiction as in case 8.2.

Remark that in cases 8.2 and 8.3 we disregard the last part of the decomposition, whether it is 1:26 (for case #17) or 1:2 + 1:6 (for case #16).

In the second part of this chapter, we are looking at all the embeddings of $4K_{2,3} + K_{2,2}$ in K_9, such that all the four $K_{2,3}$ have pairs which are mutually disjoint, and all the decomposition satisfy the established restrictions. For case #12, we disregard all those embeddings for which

$K_9 \backslash (4K_{2,3} + K_{2,2})$ has an isolated edge, or has a connected component having an odd number of edges. After deleting isomorphic duplicates, we are left with the following cases.

8.4

		x	w	y	z	C		D	B	A
12:345	1	1		-1	-1		-1			
34:567	2	1		-1			1	-1		
67:125	3	-1	1		-1					-1
89:135	4	-1	1						-1	1
68:79	5 *	-1	-1	-1	-1					
2:19	6		-1	1		1		-1		
8:26	7		-1	1		-1			-1	
9:47	8				1	1		1		-1
4:38	9				1	-1	-1		1	

R.3: $a(3) < 0$
R.4: $b(3) < 0$
R.7: $c(3) > 0$
R.6: $d(3) > 0$
contradiction in R.8.

Another decomposition is obtained from the one in 8.4 by replacing the four $K_{1,2}$ with 2:18 + 8:46 + 4:39 + 9:27. This one is isomorphic to 8.4 by the permutation (67)(89), leading to a similar contradiction, therefore it is omitted.

8.5

		x	y	w	z	C	D		B	A
12:345	1	1		-1	-1		-1			
34:567	2	1		-1			1	-1		
67:125	3	-1	1			1		-1		
89:156	4	-1	1			-1			-1	
38:49	5 *	-1	-1	-1	-1					
2:19	6		-1	1	-1					-1
8:23	7		-1	1					-1	1
9:47	8				1	1		1		-1
7:68	9				1	-1	-1		1	

R.6: $a(2) < 0$
R.7: $b(2) < 0$
R.4: $c(2) > 0$
R.1: $d(2) > 0$
contradiction in R.9.

Another decomposition is obtained from the one in 8.5 by replacing the four $K_{1,2}$ with 2:18 + 8:37 + 7:69 + 9:24. This one is isomorphic to 8.5 by the permutation (34)(89), leading to a similar contradiction, therefore it is omitted.

8.6

12:345
34:567
67:125
89:356
18:29
9:24
7:69
8:17
4:38

	x	y	w	z	C	D	A	B	
1	1		-1		1			-1	
2	1		-1		-1	-1			
3	-1	1		-1					-1
4	-1	1				-1			1
5*	-1	-1	-1	-1					
6		-1	1	-1			-1		
7		-1	1				1	-1	
8				1	1			1	-1
9				1	-1	1	-1		

R.6: $a(3) < 0$
R.7: $b(3) < 0$
R.1: $c(3) < 0$
R.2: $d(3) > 0$
contradiction in R.9.

Another decomposition is obtained from the one in 8.6 by replacing the four $K_{1,2}$ with 9:27 + 7:68 + 8:14 + 4:39. This one is isomorphic to 8.6 by the permutation (12)(89), leading to a similar contradiction, hence it is omitted.

The first part of the decomposition in the next six cases (8.7 - 8.12) leads to two different completions. We present both of them, as cases i and ii, and we split accordingly the corresponding matrices into three parts. The leftmost six columns are common to the two cases, followed by the next three columns for case i and the rightmost three columns belonging to case ii. The proof has a common beginning and occasionally splits at the last few steps, as follows.

8.7

12:345
34:567
67:158
89:135
26:79
2:16

case i	case ii
8:29	8:24
4:38	9:78
9:47	4:39

							case i			case ii		
	w	x	y	z	B	A	C	D	E	C	E	D
1	1		-1	-1		-1						
2	1				1	1	-1			-1		
3	-1	1		-1				-1				-1
4	-1	1						1	-1	-1		1
5*	-1	-1	-1	-1								
6		-1	1		1	-1						
7		-1	1		-1				-1		-1	
8			-1	1			1	-1		1	-1	
9				1	-1		-1		1		1	-1

R.1: $a(2) < 0$
R.6: $b(2) < 0$
R.2: $c(2) < 0$
R.3: $d(2) > 0$
R.7: $e(2) > 0$
case i :
contradiction in R.9.
case ii :
contradiction in R.8.

8.8

12:345
34:567
67:158
89:135
29:78
4:38

case i	case ii
2:19	2:17
7:26	9:24
9:47	7:69

	x	y	A	z	B	w	C (case i)	E (case i)	(case i)	D (case ii)	(case ii)	F (case ii)
1	1		-1	-1			-1			-1		
2	1				1		1	-1		1	-1	
3 *	-1	-1		-1		-1						
4	-1	-1				1			-1		-1	
5	-1	1	-1	-1								
6		1	1		-1			-1				-1
7		1	1					1	-1	-1		1
8			-1	1	-1	-1						
9				1	1		-1		1		1	-1

R.5: $a(1) < 0$
R.8: $b(1) > 0$
R.1: $c(1) > 0$
& $d(1) > 0$
R.2: $e(1) > 0$
case i :
contradiction in R.6.
R.6: $f(1) < 0$
case ii :
contradiction in R.7.

8.9

12:345
34:567
67:158
89:235
16:29
7:26

case i	case ii
4:38	4:39
9:47	8:14
8:19	9:78

	x	w	y	z	A	B	C (case i)	D (case i)	E (case i)	C (case ii)	E (case ii)	D (case ii)
1	1		-1		1				-1		-1	
2	1			-1	-1	-1						
3	-1	1		-1			-1			-1		
4	-1	1					1	-1		1	-1	
5 *	-1	-1	-1	-1								
6		-1	1		1	-1						
7		-1	1			1		-1				-1
8			-1	1			-1		1		1	-1
9				1	-1			1	-1	-1		1

If $a(2) > 0$
R.2: $b(2) < 0$
contradiction in R.6;
hence $a(2) \leq 0$
R.2: $b(2) > 0$
R.3: $c(2) < 0$
R.1: $e(2) < 0$
R.7: $d(2) > 0$
case i:
contradiction in R.4;
case ii:
contradiction in R.8.

8.10

12:345
34:567
68:125
79:156
37:89
2:17

case i	case ii
4:38	4:39
9:24	8:46
8:69	9:28

	w	x	y	z	B	A	E (case i)	D (case i)	C (case i)	E (case ii)	C (case ii)	D (case ii)
1	1		-1	-1		-1						
2	1		-1			1		-1				-1
3	-1	1			1		-1			-1		
4	-1	1					1	-1		1	-1	
5 *	-1	-1	-1	-1								
6		-1	1	-1					-1		-1	
7		-1		1	1	-1						
8			1		-1		-1		1		1	-1
9				1	-1			1	-1	-1		1

R.1: $a(2) < 0$
R.7: $b(2) < 0$
R.6: $c(2) > 0$
R.2: $d(2) < 0$
case i:
R.8: $e(2) > 0$
contradiction in R.4;
case ii:
contradiction in R.8.

8.11

12:345
34:567
68:125
79:156
38:49
8:36

case i	case ii
9:24	9:47
7:89	2:19
2:17	7:28

	x	y	w	z	B	A	C (case i)	D (case i)	E (case i)	C (case ii)	E (case ii)	D (case ii)
1	1		-1	-1					-1		-1	
2	1		-1				-1		1		1	-1
3	-1	1			1	-1						
4	-1	1			-1		-1			-1		
5 *	-1	-1	-1	-1								
6		-1	1	-1		-1						
7		-1		1				1	-1	-1		1
8			1		1	1		-1				-1
9				1	-1		1	-1		1	-1	

R.6: $a(2) < 0$
R.3: $b(2) < 0$
R.4: $c(2) > 0$
R.8: $d(2) < 0$
R.1: $e(2) > 0$
contradiction in R.7.

8.12

12:345
34:567
68:125
79:158
36:89
4:38

case i	case ii
7:26	7:69
9:47	2:17
2:19	9:24

	x	y	B	C	A	w	D (case i)	z (case i)	E (case i)	D (case ii)	E (case ii)	z (case ii)
1	1		-1	-1					-1		-1	
2	1		-1				-1		1		1	-1
3	-1	-1			1	1						
4*	-1	-1				-1		-1				-1
5	-1	1	-1	-1								
6		1	1		1		-1			-1		
7		1		1			1	-1		1	-1	
8			1	-1	-1	1						
9				1	-1			1	-1	-1		1

R.3: $a(2) > 0$
if $b(2) < 0$
R.8: $c(2) < 0$
contradiction in R.5;
hence $b(2) \geq 0$
R.6: $d(2) > 0$
R.2: $e(2) > 0$
R.1: $c(2) < 0$
contradiction in R.9.

This completes the proof that case #12 is impossible.

In the last part of this chapter, it remains to consider embeddings of $4K_{2,3} + 2K_{2,2}$ in K_9, subject to the established restrictions, in which the four $K_{2,3}$'s have mutually disjoint pairs. The remaining four edges of $K_9 \backslash (4K_{2,3} + 2K_{2,2})$ are to be decomposed into four $K_{1,1}$ (for case #16), to $K_{1,2} + 2K_{1,1}$ (for case #17) or to two $K_{1,2}$ (for case #18).

Similar to the previous cases, we let the machine find all the combinatorially different embeddings of the type under consideration. Avoiding isomorphic duplicates, we get just four such decompositions. The remaining graph in these decompositions contains four mutually disjoint edges, hence it is related only to case #16. The cases are as follows.

8.13

12:345
34:567
67:125
89:135
24:89
68:79
1:2
3:4
6:8
7:9

	x	w	y	z	B		C	A		
1	1		-1	-1			1			
2	1		-1		1		-1			
3	-1	1		-1				1		
4	-1	1			1			-1		
5 *	-1	-1	-1	-1						
6		-1	1			1			1	
7		-1	1			-1				1
8				1	-1	1			-1	
9				1	-1	-1				-1

R.3: a(1) > 0
R.4: b(1) > 0
R.2: c(1) > 0
contradiction in R.1.

8.14

12:345
34:567
67:125
89:156
27:89
38:49
1:2
6:7
3:8
4:9

	x	y	w	z	B		C	A		
1	1		-1	-1			1			
2	1		-1		1		-1			
3	-1	1				1			1	
4	-1	1				-1				1
5 *	-1	-1	-1	-1						
6		-1	1	-1				1		
7		-1	1		1			-1		
8				1	-1	1			-1	
9				1	-1	-1				-1

R.6: a(1) > 0
R.7: b(1) > 0
R.2: c(1) > 0
contradiction in R.1.

8.15

12:345
34:567
67:125
89:356
18:29
47:89
1:8
2:9
3:4
6:7

	x	y	w	z		B			C	A
1	1		-1		1		1			
2	1		-1		-1			1		
3	-1	1		-1					1	
4	-1	1				1			-1	
5 *	-1	-1	-1	-1						
6		-1	1	-1						1
7		-1	1			1				-1
8				1	1	-1	-1			
9				1	-1	-1		-1		

R.6: a(2) > 0
R.7: b(2) > 0
R.4: c(2) > 0
contradiction in R.3.

8.16

12:345
34:567
67:589
89:125
16:27
38:49
1:6
2:7
3:8
4:9

	x	B	C	y	w		z	A		
1 *	1			1	1		1			
2	1			1	-1			1		
3	-1	1				1			1	
4	-1	1				-1				1
5	-1	-1	-1	1						
6		-1	1		1		-1			
7		-1	1		-1			-1		
8			-1	-1		1			-1	
9			-1	-1		-1				-1

R.2: a(3) < 0
if b(3) > 0
R.6: c(3) > 0
contradiction in R.5;
hence b(3) ≤ 0
R.5: c(3) > 0
contradiction in R.7.

Therefore the cases #12, 16, 17 and #18 are all impossible.

Chapter 9

In this chapter we treat the cases #20, 21, 22 and #23.

In these last four cases, we look for embeddings of five $K_{2,3}$'s in K_9, subject to the established restrictions. The five pairs in the (five) $K_{2,3}$'s must contain two intersecting pairs. Thus we can assume, without loss of generality, that $K_{2,3}(1) = 12:345$ and $K_{2,3}(2)$ is one of 12:678, 13:678 or 16:789. We let the machine find all the embeddings under consideration. In each case we get a contradiction, as follows.

9.1

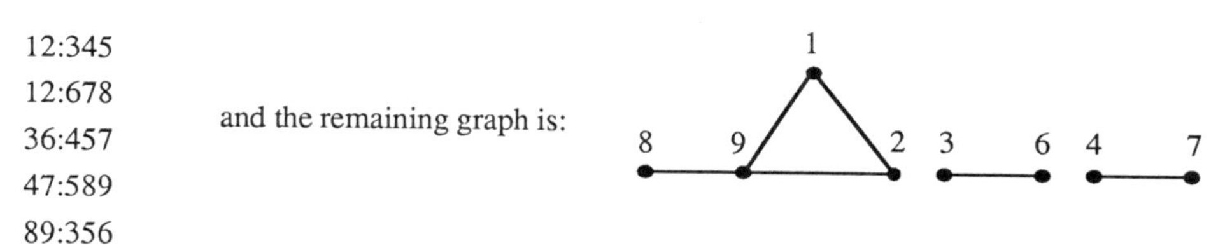

It suffices to treat the matrix related to just $5K_{2,3} + 3{:}6{+}4{:}7$, as follows.

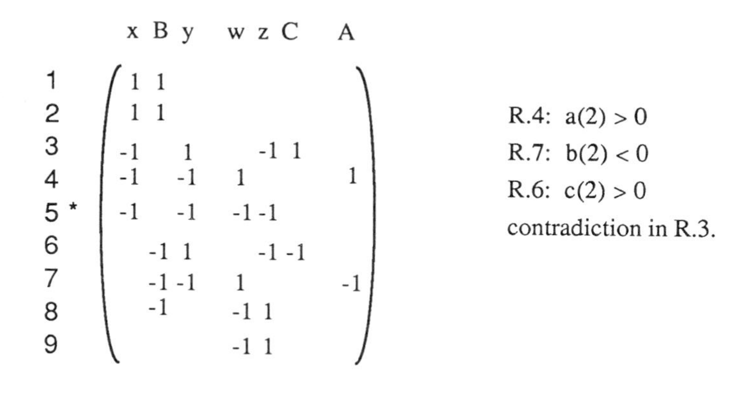

9.2

12:345
12:678
36:479
47:589
58:369

and the remaining graph is:

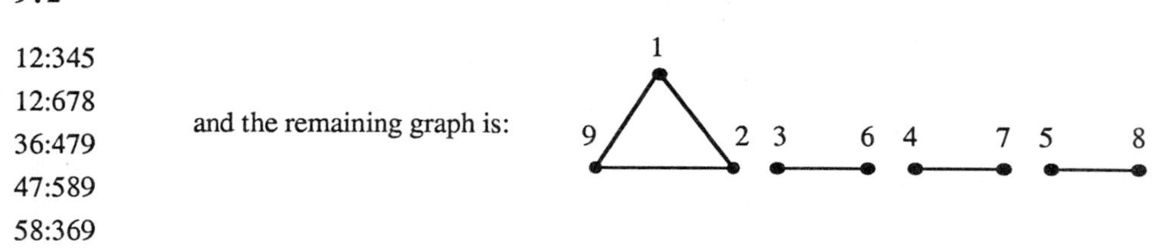

In this case the remaining graph $K_9\backslash 5K_{2,3}$ contains no two edge-disjoint $K_{1,2}$'s, therefore it can only be decomposed into $K_{1,2} + 4K_{1,1}$ for case #20. If $K_{1,2} = 9{:}12$, then $m(1, 2) = 3$, in contradiction with Lemma 9. If $K_{1,2} = 1{:}29$ $(2{:}19)$, then the vertex 1 (2, respectively) is of type (3, 3, 2, 0), hence $\alpha \geq 1$; this is a contradiction since in case #20 $x_{0,1} = 0 = \alpha$. Therefore this case is impossible.

9.3

12:345
12:678
36:479
49:578
58:367

and the remaining graph is:

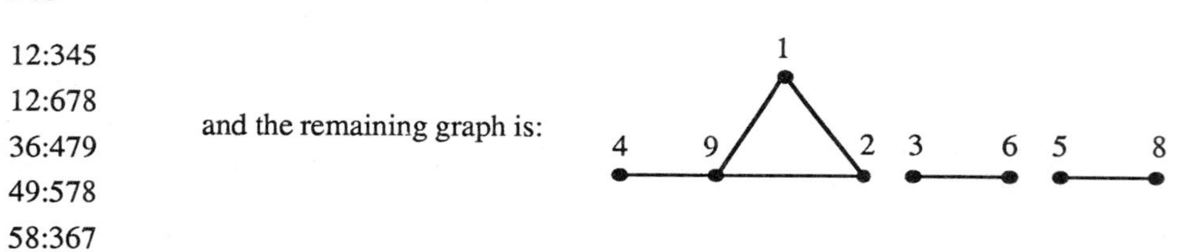

The matrix corresponding to the $5K_{2,3} + 3{:}6 + 5{:}8$ is as follows.

	B	x	z	y	w	C	A
1	1	1					
2	1	1					
3	-1		1		-1	1	
4	-1		-1	1			
5	-1			-1	1		1
6		-1	1		-1	-1	
7 *		-1	-1	-1	-1		
8		-1		-1	1		-1
9			-1	1			

R.8: $a(3) < 0$
R.5: $b(3) < 0$
R.3: $c(3) < 0$
contradiction in R.6.

9.4

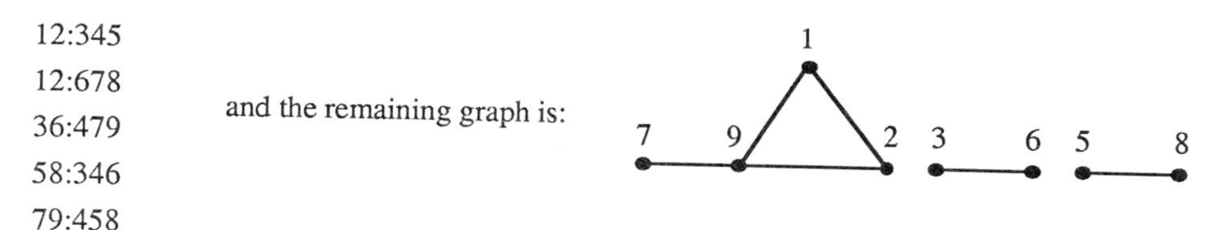

12:345
12:678
36:479
58:346
79:458

and the remaining graph is:

The matrix corresponding to the $5K_{2,3} + 3{:}6 + 5{:}8$ is as follows.

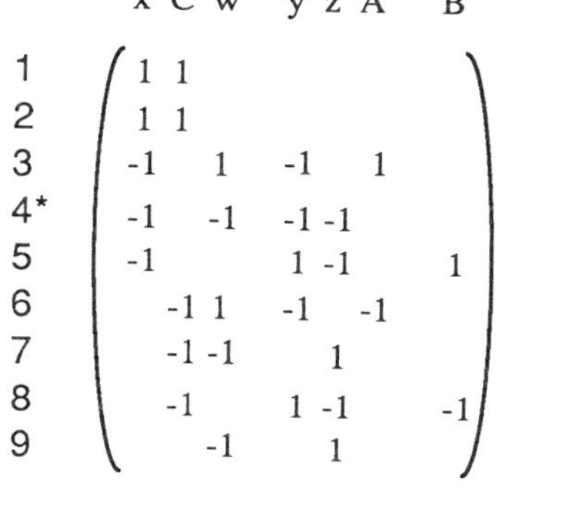

	x	C	w	y	z	A	B
1	1	1					
2	1	1					
3	-1		1	-1		1	
4*	-1		-1	-1	-1		
5	-1			1	-1		1
6		-1	1	-1		-1	
7		-1	-1		1		
8		-1		1	-1		-1
9			-1		1		

R.3: $a(2) > 0$
R.5: $b(2) < 0$
R.8: $c(2) > 0$
contradiction in R.6.

9.5

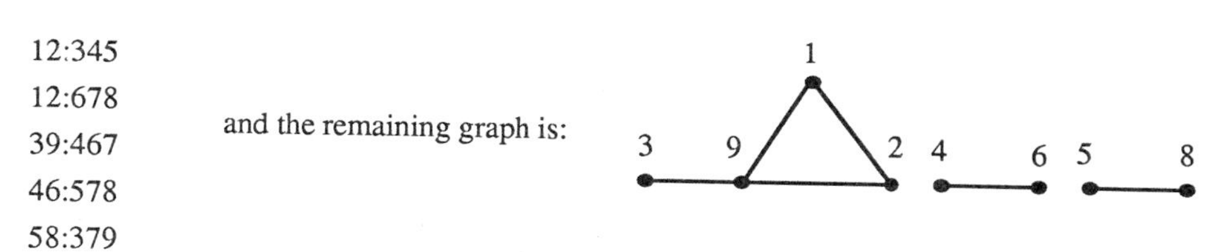

12:345
12:678
39:467
46:578
58:379

and the remaining graph is:

The matrix corresponding to the $5K_{2,3} + 4{:}6 + 5{:}8$ is as follows.

	B	x	y	w	z	A	C
1	1	1					
2	1	1					
3	-1		1		-1		
4	-1		-1	1		1	
5	-1			-1	1		1
6		-1	-1	1		-1	
7*		-1	-1	-1	-1		
8		-1		-1	1		-1
9			1		-1		

R.6: $a(3) < 0$
R.4: $b(3) < 0$
R.5: $c(3) < 0$
contradiction in R.8.

This completes the proof that cases #20, 21, 22 and #23 are impossible.

This also completes the proof of the Main Theorem, thus

there exist no neighborly families consisting of nine tetrahedra.

In the early stages of our computer search we found out, by using only graph theoretic reasons, like Lemma 10, that the folowing holds.

Theorem 5: The graph K_9 does not have decompositions of the following types:

$$K_9 \neq 3K_{2,3} + 4K_{2,2} + K_{1,2} ,$$

$$K_9 \neq 3K_{2,3} + 4K_{2,2} + 2K_{1,1} ,$$

$$K_9 \neq 4K_{2,3} + 2K_{2,2} + 2K_{1,2} ,$$

and $$K_9 \neq 5K_{2,3} + 3K_{1,2} .$$

We omit the details of the proof-by-computer, for obvious reasons. Theorem 5 implies that cases #10, 9, 18 and #22 of Table 1 are impossible, because the corresponding decompositions of K_9 are impossible.

Chapter 10

In this chapter we introduce the notion of nearly-neighborliness, and treat neighborly families consisting of eight tetrahedra.

A family of convex d-polytopes F in E^d is called *nearly-neighborly* if for every two members of F there exists a hyperplane which separates them and contains a facet of each one of them. This notion was introduced by us in our first result [17] on neighborly families of convex polytopes. Clearly every neighborly family is also a nearly-neighborly family.

Following [17], let f(d, k) (g(d, k)) denote the maximum number of members in a neighborly (nearly-neighborly, respectively) family of convex d-polytopes in E^d, in which every member has at most k facets.

Our Main Theorem can be stated as follows: *f(3, 4) = 8.*

The argument which Bagemihl used in [1] to show that there can be at most seventeen neighborly tatrahedra (i.e., f(3, 4) $\leq$ 17), involves a maximum nearly-neighborly set of triangles in the plane, and it can be stated as follows (see also [17] and [14]).

Lemma 11: g(2, 3) = 4.

We proved in [17] that f(d, k) $\leq$ g(d, k) $\leq$ (3/2)k!, and Perles [12] reduced the upper bound to g(d, k) $\leq 2^k$. A refinement of Perles' reasons leads to a reduction of the upper bound by 1, due to M. Markert (see [14]), showing that f(d, k) $\leq$ g(d, k) $\leq 2^k$ - 1. We reproduce these arguments here, as follows.

Theorem 5: $g(d, k) \leq 2^k - 1$, for every $d \geq 2$ and every $k \geq d+1$.

Proof of Theorem 5: Let $F = \{P_1, ..., P_n\}$ be a nearly-neighborly family consisting of n d-polytopes in E^d, and let $\{H_1, ..., H_t\}$ be the collection of hyperplanes in E^d containing a facet of at least one member of F. Let B(F) be the Baston matrix of F, of size nxt, defined in the same way as it was defined for neighborly families of convex polytopes. The matrix B(F) has the following property (compare with (2)).

(2*) For every pair of row-indices i and j, there exists *at least one* column-index k, such that

$$b_{ik} \cdot b_{jk} = -1,$$

because two members of F are separated by at least one, but possibly more than one, hyperplane containing a facet of each one of them.

For simplicity, suppose that all the members of F have precisely k facets. The following holds (compare with (1)).

(1*) Each row of B(F) has precisely k nonzero terms.

Let $C(F) = (c_{i,j})$ be the matrix of size $(n \cdot 2^{t-k})$xt, obtained from B(F) as follows: replace every row with 2^{t-k} rows, in which all the $t - k$ zeros are replaced by +1 or -1, in all the 2^{t-k} different ways.

Following Perles' reasoning, all the rows of C(F) are distinct, due to the construction and to the nearly-neighborliness of F. As C(F) has t columns, and its terms are only +1 and -1, it follows that $n \cdot 2^{t-k} \leq 2^t$, implying that $n \leq 2^k$.

To get $n \leq 2^k - 1$, we argue as follows. The i-th row of B(F) represents P_i, in the form $P_i = \cap H_j^*$, where H_j^* means H_j^+, if $b_{i,j} = 1$, it means H_j^- if $b_{i,j} = -1$, and it means that the term does not appear at all in the intersection if $b_{i,j} = 0$. Correspondingly, the i-th row of C(F) represents a set Q_i of the form $Q_i = \cap H_j^*$, where H_j^* means H_j^+ or H_j^-, depending on whether $c_{i,j} = 1$ or $c_{i,j} = -1$. Clearly, all the Q_i's are convex polyhedra; moreover, each one of them is contained in one of the P_j's, hence it is *bounded*. It follows that all the *unbounded* regions of E^d, determined by these t hyperplanes, are not presented by any one of the rows of C(F). Therefore C(F) contains less than 2^t rows. Hence $n \cdot 2^{t-k} \leq 2^t - 1$, implying that $n \leq 2^k - 1$.

Of course, the same arguments show that C(F) has much less than 2^t rows, by a polynomial in t of degree d, but we cannot use it to get a better upper bound for n.

This completes the proof of Theorem 5.

Considering nearly-neighborly families of four triangles in the plane, as studied extensively by Simon [14], one can get the following.

Lemma 12: If four triangles in the plane form a nearly-neighborly family, then no line, which contains an edge of each one of the triangles, separates one of the triangles from the remaining three.

Lemma 12 and the notion of nearly-neighborly are used in our study of neighborly families of eight tetrahedra.

So far, only two examples of neighborly families consisting of eight tetrahedra were given in the literature. These are the two families mentioned earlier, see Figure 1 and Figure 2. One can get, yet, a third neighborly family consisting of eight tetrahedra, by taking two halves, one from each one of these two examples, and combine them together, as shown in Figure 4.

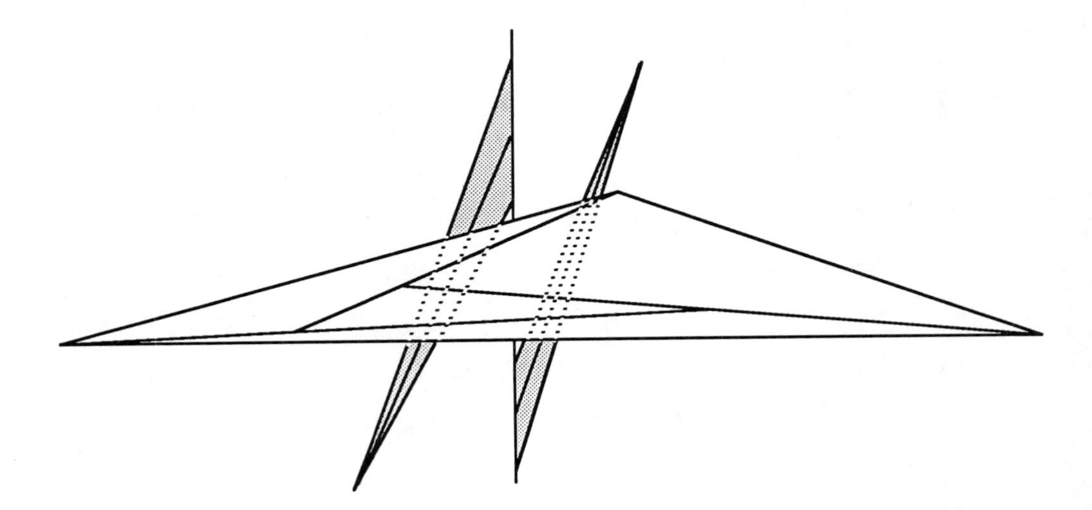

Figure 4: Another example of eight neighborly tetrahedra, in which four are taken from the Bagemihl's example and the other four are taken from the Wilson-Zaks' example.

However, we know of no other examples of neighborly families consisting of eight tetrahedra. This leads us to the following problems.

Problem 1: *Are there any other neighborly families, consisting of eight tetrahedra, besides the aforementioned three?*

In particular, we are interested in the following.

Problem 2: *Is it true that for every neighborly family, consisting of eight tetrahedra, there exists a plane which contains a facet of each one of the tetrahedra?*

Let F be a neighborly family in E^3, consisting of eight tetrahedra. The Baston matrix B(F) of F will satisfy the following relations, in addition to (1) and (2).

(3*) $\quad x_{i,j} > 0 \quad$ implies $j \leq 4$.

(7*) $\quad \sum_{i,j} (i+j)x_{i,j} = 32 \quad (= 8x4)$.

(8*) $\quad \sum_{i,j} ijx_{i,j} = 28 \quad (= \binom{8}{2})$.

(9*) $\quad K_8 = \sum_{1\leq i\leq j\leq 4} x_{i,j} K_{i,j}$.

(10*) $\quad \sum_{1\leq i\leq j\leq 4} x_{i,j} \geq 7$.

Property (3*) is implied by Lemma 11 (see [17]), and properties (7*), (8*), (9*) and (10*) are the analogues of (7), (8), (9) and (10), respectively, when considering neighborly families consisting of eight tetrahedra. To try and solve Problem 1, one may wish to consider all the solutions of the Diophantine system (3*, 7*, 8*, 10*), for which there exists a decomposition of K_8 of the form (9*). However, the number of solutions is about five hundred, and we could not conclude much from them.

The situation is somewhat simpler if one assumes that Problem 2 has an affirmative answer, as follows.

Theorem 6: If F is a neighborly family, consisting of eight tetrahedra, for which there exists a plane which contains a facet of each member of F (i.e., $x_{4,4} \geq 1$), then $x_{4,4} = 1$, $x_{i,j} = 0$ for all i and j satisfying $i + j \geq 5$, $x_{2,2} + x_{1,3} \leq 2$, and there are 22 possible solutions to the system (3*, 7*, 8*, 10*), as given in Table 2.

Case #	$x_{0,1}$	$x_{1,1}$	$x_{1,2}$	$x_{1,3}$	$x_{2,2}$	$x_{4,4}$	Remarks
1.	8	4	0	0	2	1	The Wilson-Zaks example (see Figure 2).
2.	7	3	1	1	1	1	
3.	6	5	0	1	1	1	
4.	7	2	3	0	1	1	The third example (see Figure 3).
5.	6	4	2	0	1	1	
6.	5	6	1	0	1	1	
7.	4	8	0	0	1	1	
8.	6	2	2	2	0	1	
9.	5	4	1	2	0	1	
10.	4	6	0	2	0	1	
11.	6	1	4	1	0	1	
12.	5	3	3	1	0	1	
13.	4	5	2	1	0	1	
14.	3	7	1	1	0	1	
15.	2	9	0	1	0	1	
16.	6	0	6	0	0	1	Bagemihl's example (see Figure 1).
17.	5	2	5	0	0	1	
18.	4	4	4	0	0	1	
19.	3	6	3	0	0	1	
20.	2	8	2	0	0	1	
21.	1	10	1	0	0	1	
22.	0	12	0	0	0	1	

Table 2. The 22 solutions of the equations (3*, 7*, 8*, 10*).

Proof of Theorem 6: Let F be a neighborly family, consisting of eight tetrahedra, and let H be the plane which contains a facet of each one of the P_i's, separating them into two quadruples. By property (9*), K_8 has a decomposition into complete bipartite graphs, including a $K_{4,4}$. The graph $K_8 \backslash K_{4,4}$ is the disjoint union of two K_4's. Therefore every member of the rest of the decomposition contains at most four vertices, hence $x_{4,4} = 1$ and $x_{i,j} = 0$ for all i and j satisfying $i + j \geq 5$.

Each copy of K_4 can accommodate one of $K_{2,2}$ or $K_{1,3}$, hence $x_{2,2} + x_{1,3} \leq 2$.

This completes the proof of Theorem 6.

It will be interesting to settle the two problems mentioned in this chapter, and to see if Table 2 might be of some use.

The problem of determining the maximum number of members in a neighborly family of d-simplices in E^d, $f(d, d+1)$, for $d \geq 4$, is as follows. $2^d \leq f(d, d+1) \leq 2^{d+1} - 1$, where the lower bound is due to [18] and the upper bound was shown here as Theorem 5, due to Perles and Markert.

It has been repeatedly conjectured (see [5, 8, 12, 17]) that $f(d, d+1) = 2^d$, for all $d \geq 3$. The amount of details it took to establish the three dimensional case does not leave much of a hope for the higher dimensional cases. The situation in four dimensions is: $16 \leq f(4, 5) \leq 31$, and it is conjectured that $f(4, 5) = 16$.

In a paraphrase of [10], we raise the following.

Problem 3: *Can seventeen 4-simplices form a neighborly family in E^4 ?*

Chapter 11

We conclude our paper by presenting, from [6], the proof that $g(3, 4) \leq 14$, as stated in the next theorem.

Theorem 7: *A nearly-neighborly family in E^3 contains at most fourteen tetrahedra.*

Our proof of Theorem 7, that $g(3, 4) \leq 14$, is similar to the one given in [23], where $g(3, 4) \leq 15$ was established; it, too, uses the idea of Tverberg's [16] proof of the Graham-Pollak Theorem, extended to a few cases where a multigraph is decomposed into complete bipartite graphs.

We need the following four Lemmas, where the following notations are used: the vertex set $V(K_{15})$ of K_{15} is taken as $\{1, ..., 15\}$. If for some multigraph H, $\{1, ..., 15\} \supset V(H)$, then $K_{15} \oplus H$ will denote the edge-disjoint union of K_{15} and H, where all the (multiple) edges of H are added as multiple edges in $K_{15} \oplus H$. The vertex-disjoint union of two graphs G and H will be denoted by $G +: H$.

Lemma 13: If H is a multigraph, $\{1, ..., 15\} \supset V(H)$ and if H minus *five* vertices has at most one edge, then $K_{15} \oplus H$ is not the edge-disjoint union of *eight* complete bipartite graphs.

Lemma 14: If H is a multigraph, $\{1, ..., 15\} \supset V(H)$ and if H minus *two* vertices consists of *five* independent edges, then

$$K_{15} \oplus H \neq \sum_{i=1}^{8} K_{A_i, B_i}$$

where $|A_1|, |B_1| \leq 4$ and $|A_1 \cup B_1| \geq 6$.

Lemma 15: If H is a multigraph, $\{1, ..., 15\} \supset V(H)$ and if H minus *one* vertex consists of *six* independent edges, then

$$K_{15} \oplus H \neq \sum_{i=1}^{8} K_{A_i, B_i}, \quad 4 \leq |A_1| \leq 5.$$

Lemma 16: the following hold:

(i) The graph $K_{15} \oplus (C_5 +: C_3 +: C_3)$ has no decompositions of the form $K_{2,2} + 7K_{4,4}$,

(ii) The graph $K_{15} \oplus (C_2 +: C_3 +: C_3 +: C_3)$ has no decompositions of the form $K_{2,2} + 7K_{4,4}$ and

(iii) The graph $K_{15} \oplus (C_3$ +: six independent edges) has no decompositions of the form $2K_{3,3} + 6K_{4,4}$.

Proof of Lemma 13: Let H be any multigraph, $\{1, ..., 15\} \supset V(H)$, such that $H \setminus \{11, 12, 13, 14, 15\}$ has just the edge (1,2) (or none), and suppose that

$$K_{15} \oplus H = \sum_{i=1}^{8} K_{A_i, B_i}.$$

Let us consider the following system of linear homogeneous equations:

$$\sum_{i \in A_j} x_i = 0, \quad 1 \leq j \leq 8$$

$$x_i = 0 \qquad \text{for all } i,\ 11 \leq i \leq 15,$$

and

$$\sum_{i=1}^{15} x_i = 0$$

with the variables $x_1, x_2, ..., x_{15}$.

Squaring the last equation we get

$$0 = \left(\sum_{i=1}^{15} x_i\right)^2 = \sum_{i=1}^{15} x_i^2 + 2 \sum_{i<j} x_i x_j$$

$$= \sum_{i=1}^{15} x_i^2 + 2\left[\sum_{j=1}^{8}\left(\sum_{i\in A_j} x_i\right)\left(\sum_{i\in B_j} x_i\right) - \sum_{\substack{i<j \\ (i,j)\in H}} x_i x_j\right]$$

$$= \sum_{i=1}^{15} x_i^2 - 2x_1x_2$$

$$= \sum_{i\geq 3} x_i^2 + (x_1 - x_2)^2 .$$

We have made use of the edge-disjoint decomposition

$$K_{15} \oplus H = \sum_{i=1}^{8} K_{A_i, B_i}$$

and the eight equations of the form

$$\sum_{i\in A_j} x_i = 0$$

which nullify the main term, as well as making use of the equations $x_i = 0$ for $11 \leq i \leq 15$, which implies that

$$\sum_{\substack{i<j \\ (i,j)\in H}} x_i x_j = x_1 x_2 .$$

Thus it follows that $x_1 = x_2$ and $x_i = 0$ for all $i \geq 3$. However, as

$$\sum_{i=1}^{15} x_i = 0 ,$$

it follows that $x_1 = x_2 = 0$, hence the system has only the trivial solution. This is impossible, since there are only fourteen equations involving the fifteen variables.

This completes the proof of Lemma 13.

Proof of Lemma 14 (see Theorem 3 in [23]: Let H be a multigraph, $\{1, ..., 15\} \supset V(H)$, such that $H\setminus\{14, 15\}$ consists of the edges $(1, 2), (3, 4), (5, 6), (7,8)$ and $(9, 10)$, and suppose that

$$K_{15} \oplus H = \sum_{i=1}^{8} K_{A_i, B_i}$$

where $|A_1|, |B_1| \leq 4$ and $|A_1 \cup B_1| \geq 6$.

Consider the following system:

$$x_{14} = x_{15} = 0$$

$$\sum_{i \in A_j} x_i = 0, \quad 1 \leq j \leq 8$$

and

$$\sum_{i=1}^{15} x_i = 0 .$$

Using the idea of the proof of the previous lemma, we get

$$x_1 = x_2 = a_1$$

$$x_3 = x_4 = a_2$$

$$x_5 = x_6 = a_3$$

$$x_7 = x_8 = a_4$$

$$x_9 = x_{10}$$

and $x_i = 0$ for all i, $11 \leq i \leq 15$.

The equation

$$\sum_{i=1}^{15} x_i = 0$$

implies that $x_9 = x_{10} = - a_1 - a_2 - a_3 - a_4$, hence the dimension of the solution space is at most four.

As 15 = the number of variables = rank of the system + dimension of the solution space ≤ 11 + 4 = 15, it follows that we do have, indeed, a general solution, involving four parameters.

Following the proof of Theorem 3 in [23], we conclude that for all j, $1 \leq j \leq 8$, either $|A_j| \geq 5$, if $A_j \cap \{1, ..., 10\} \neq \emptyset$, or else $\{11, ..., 15\} \supseteq A_j$ (implying that $|A_j| \leq 5$).

The set A_1 satisfies $|A_1| \leq 4$, hence it follows that $\{11, ..., 15\} \supseteq A_1$.

Repeating the same argument, with the B_j's replacing the A_j's, we conclude that $\{11, ..., 15\} \supseteq B_1$. However, $|A_1 \cup B_1| \geq 6$ and $A_1 \cap B_1 = \emptyset$ contradict $\{11, ..., 15\} \supseteq A_1, B_1$.

This completes he proof of Lemma 14.

The particular case of Lemma 14 that we need is where $|A_1| = |B_1| = 4$, as stated in the following lemma.

Lemma 14*: If H is a multigraph, $\{1, ..., 15\} \supset V(H)$ and if H minus *two* vertices consists of *five* independent edges, then

$$K_{15} \oplus H \neq K_{4,4} + \sum_{i=2}^{8} K_{A_i, B_i} .$$

Proof of Lemma 15: Let H be a multigraph , $\{1, ..., 15\} \supset V(H)$, such that

$H \setminus \{15\} = \cup_{1 \le i \le 6} (2i - 1, 2i)$ and suppose that

$$K_{15} \oplus H = \sum_{i=1}^{8} K_{A_i, B_i}, \quad 4 \le |A_1| \le 5.$$

Consider the following system:

$$x_{15} = 0$$

$$\sum_{i \in A_j} x_i = 0, \quad 1 \le j \le 8$$

and

$$\sum_{i=1}^{15} x_i = 0.$$

It follows that $x_{2i-1} = x_{2i}$ for $1 \le i \le 6$ and $x_i = 0$ for all i, $13 \le i \le 15$. Setting $x_{2i-1} = x_{2i} = a_i$, $1 \le i \le 5$, it follows from

$$\sum_{i=1}^{15} x_i = 0$$

that $x_{11} = x_{12} = -a_1 - a_2 \ldots - a_5$, hence the dimension of the solution space is at most five.

Again, 15 = the number of variables = rank of the system + dimension of the solution space $\le 10 + 5 = 15$, hence the dimension of the solution space is equal to five. Thus, for all j, $1 \le j \le 8$, either $|A_j| \ge 6$ in case $A_j \cap \{1, \ldots, 12\} \ne \emptyset$, or else $\{13, 14, 15\} \supseteq A_j$, implying that $|A_j| \le 3$. Thus the given condition $4 \le |A_1| \le 5$ leads to a contradiction.

This completes the proof of Lemma 15.

We need the following particular case of Lemma 15.

Lemma 15*: If H is a multigraph, $\{1, \ldots, 15\} \supset V(H)$ and if H minus *one* vertex consists of *six* independent edges, then

$$K_{15} \oplus H \neq K_{4,4} + \sum_{i=2}^{8} K_{A_i, B_i} .$$

Lemmas 13, 14 and 15 can be easily generalized.

Proof of Lemma 16: To show case (i) and (ii), Let D be either a $C_5 = (7, 8) \cup (8, 9) \cup (9, 10) \cup (10, 11) \cup (11, 7)$ or a $C_3 +: C_2 = (7, 8) \cup (8, 9) \cup (9, 7) + 2(10, 11)$, and let $H = D +: C_3 +: C_3$, where $C_3 +: C_3 = (1, 2) \cup (2, 3) \cup (3, 1) + (4, 5) \cup (5, 6) \cup (6, 4)$.

Suppose that

$$K_{15} \oplus H = \sum_{i=1}^{8} K_{A_i, B_i} ,$$

where $|A_1| = |B_1| = 2$ and $|A_i| = |B_i| = 4$ for all i, $2 \leq i \leq 8$.

Claim. No edge (r,s) of C_3 in H is such that $r, s \in A_j$ (or $r, s \in B_j$), $2 \leq j \leq 8$.

Proof of the Claim: Suppose, say, that $1, 2 \in A_2$. Consider the following system:

$$x_3 = x_6 = x_9 = x_{11} = 0$$

$$\sum_{i \in A_j} x_i = 0 , \quad 1 \leq j \leq 8$$

and

$$\sum_{i=1}^{15} x_i = 0 .$$

It follows, as in the previous cases, that

$$x_1 = x_2 = a$$

$$x_4 = x_5 = b$$

$$x_7 = x_8 = -a - b,$$

and all the other x_i are equal to zero.

As there are 13 equations, it follows that the general solution has two parameters; however, as $1, 2 \in A_2$, and

$$\sum_{i \in A_2} x_i = 0 ,$$

it follows that $7, 8 \in A_2$, so as to nullify $x_1 + x_2 = 2a$. Likewise, $5, 6 \in A_2$, so as to nullify $x_1 + x_2 + x_7 + x_8 = -2b$. Thus $|A_2| \geq 6$, which is a contradiction.

This completes the proof of the Claim.

The only vertices in $K_{15} \oplus H$ of valences which are not congruent to zero modulu 4 are 12, 13, 14 and 15; since $K_{15} \oplus H = K_{2,2} + 7K_{4,4}$, it follows that $V(K_{2,2}) = \{12, 13, 14, 15\}$. Therefore the two disjoint copies of C_3 in $K_{15} \oplus H$ are covered by the $7K_{4,4}$.

As no edges of the C_3's are in the same side of any of the seven $K_{4,4}$'s, we may assume, without loss of generality, that $1 \in A_2 \cap A_3 \cap B_4 \cap B_6$, $2 \in B_2 \cap A_4 \cap A_5 \cap B_7$ and 3 is in $B_3 \cap B_5 \cap A_6 \cap A_7$; thus the six edges of $2C_3$ are in six of the seven $K_{4,4}$.

As the vertex 4 is connected by simple edges to the vertices 1, 2 and 3, it follows that 4 is in precisely three of the sets in $\{A_j, B_j \mid 2 \le j \le 7\}$. The vertex 4 must belong to four $K_{4,4}$, hence $4 \in A_8 \cup B_8$.

In a similar way, $5 \in A_8 \cup B_8$ and also $6 \in A_8 \cup B_8$.

It follows that some two vertices among $\{4, 5, 6\}$ are in either A_8 or B_8, contrary to the claim.

This completes the proof of cases (i) and (ii) of Lemma 16.

To show (iii), let $H = \cup_{1 \le i \le 6}(2i-1, 2i)+(13, 14)\cup(14, 15)\cup(15, 13)$, and suppose that

$$K_{15} \oplus H = \sum_{i=1}^{8} K_{A_i, B_i} ,$$

where $|A_1| = |A_2| = |B_1| = |B_2| = 3$ and $|A_j| = |B_j| = 4$ for $3 \le j \le 8$.

The three vertices 13, 14 and 15 in $K_{15} \oplus H$ have valence 16, while the rest of the vertices are 15-valent, hence it follows that $A_1 \cup B_1 \cup A_2 \cup B_2 = \{1, ..., 12\}$.

Consider the following system:

$$x_{15} = 0$$

$$\sum_{i \in A_j} x_i = \sum_{i \in B_j} x_i \qquad 1 \le j \le 8$$

and

$$\sum_{i=1}^{15} x_i = 0 .$$

Squaring the last equation, we get:

$$0 = \left(\sum_{i=1}^{15} x_i\right)^2 = \sum_{i=1}^{15} x_i^2 + 2\sum_{i<j} x_i x_j$$

$$= \sum_{i=1}^{15} x_i^2 + 2\left[\sum_{j=1}^{8}\left(\sum_{i\in A_j} x_i\right)\left(\sum_{i\in B_j} x_i\right) - \sum_{\substack{i<j\\(i,j)\in H}} x_i x_j\right]$$

$$= \sum_{i=1}^{7}(x_{2i} - x_{2i-1})^2 + 2\sum_{j=1}^{8}\left(\sum_{i\in A_j} x_i\right)^2 .$$

It follows that

$$\sum_{i\in A_j} x_i = \sum_{i\in B_j} x_i = 0 \quad \text{for all } j,\ 1 \le j \le 8,$$

and that $x_{2i-1} = x_{2i} = a_i$ for all i, $1 \le i \le 6$ and $x_{13} = x_{14} = - a_1 - a_2 - \ldots - a_6$.

If the system of the ten equations is dependent, its rank is at most nine and we have a solution involving six independent parameters. However, by taking a j such that $1 \in A_j$, it follows that one of 13 and 14 is in A_j, and hence so is also one of $2i-1$ and $2i$, for all i, $2 \le i \le 6$. Therefore $|A_j| \ge 7$, which is impossible.

The system of equations is, therefore, independent, its rank is ten and a solution of it should have five parameters. As $x_{2i-1} = x_{2i}$ for all i, $1 \le i \le 7$ and

$$\sum_{i=1}^{7} x_{2i-1} = \sum_{i=1}^{7} x_{2i} = 0 \quad ,$$

it follows that some *five* values of i are such that $x_{2i-1} = x_{2i} = a_i$ are independent parameters; say $x_{2i-1} = x_{2i} = a_i$ for $1 \le i \le 5$. It follows that $x_{11} = x_{12} = \alpha_1 a_1 + \alpha_2 a_2 + \ldots + \alpha_5 a_5$, for some reals

$\alpha_1, ..., \alpha_5$, and $x_{13} = x_{14} = -(\alpha_1 + 1)a_1 - (\alpha_2 + 1)a_2 - ... - (\alpha_5 + 1)a_5$. All this is up to possible changes in the roles of the members in the pairs $x_{2i-1} = x_{2i}$, $1 \leq i \leq 7$.

It follows from $V(2K_{3,3}) = \{1, ..., 12\}$ that one of A_1, A_2, B_1 and B_2 contains two vertices r and s, for which $r, s \leq 12$ and x_r and x_s are equal to two different parameters.

Without loss of generality, let $1,3 \in A_1$. The equation

$$\sum_{i \in A_1} x_i = 0$$

implies that $0 = x_1 + x_3 + x_k = a_1 + a_2 + (-a_1 - a_2)$; hence for some k in A_1, $x_k = -a_1 - a_2$. It follows that $x_{11} = x_{12} = -a_1 - a_2$ $(k = 11$ or $12)$, and therefore $x_{13} = x_{14} = -a_3 - a_4 - a_5$.

The set $A_1 \cup A_2 \cup B_1 \cup B_2$ contains 5; let $5 \in C$, where C is one of A_1, A_2, B_1 or B_2. $x_5 = a_3$, hence in order for

$$\sum_{i \in C} x_i = 0$$

to hold, it follows that C must contain one of the vertices $2i - 1$ and $2i$, for which $x_{2i-1} = x_{2i} = -a_3 - a_4 - a_5$, and hence it must also contain one of the indices $2i - 1$ and $2i$ for which $x_{2i-1} = x_{2i} = a_4$ and one of the indices $2j - 1$ and $2j$ for which $x_{2j-1} = x_{2j} = a_5$. Thus $|C| \geq 4$, which is a contradiction.

This completes the proof of Lemma 16.

We are ready to present the following.

Proof of Theorem 7: Suppose there exists a nearly-neighborly family $F = \{P_1, ..., P_{15}\}$

consisting of fifteen tetrahedra in E^3. Consider its Baston matrix B(F). The corresponding quantities $x_{i,j}$ have property (3*), i.e., $x_{i,j} > 0$ implies $j \leq 4$ (see [1], and also [17]). Moreover, by Corollary 3 of [23], the fact that $|F| = 15 = 2^k - 1$ (where $k = 4$ is the number of facets of every tetrahedron of F) yields the restriction $j - i \leq 1$.

Extending the notion of type to the nearly-neighborly family F in the natural way, we get that the types of members of F are of the form (p, q, r, s), $4 \geq p \geq q \geq r \geq s$, such that $14 \leq p + q + r + s \leq 16$. The only possible types are therefore (4, 4, 4, 4), (4, 4, 4, 3), (4, 4, 4, 2) and (4, 4, 3, 3). Consequently, $x_{i,j} > 0$ implies that $2 \leq i \leq j \leq 4$.

The analogue of property (7) for B(F) is obvious:

(7**) $$\sum_{i,j} (i+j)x_{i,j} = 60 \quad (=15\text{x}4).$$

The analogue of property (8) for nearly neighborly families involves an *inequality* , because two members of F may be separated by more than one hyperplane which contains facets of both of them. Thus we have

(8**) $$\sum_{i,j} ijx_{i,j} \geq 105 \quad \left(=\binom{15}{2}\right).$$

The system (7**) and (8**) in the variables $x_{2,2}$, $x_{2,3}$, $x_{3,3}$, $x_{3,4}$ and $x_{4,4}$ has nine possible solutions, presented in Table 3. The corresponding graph G(F) of F is defined in a way, similar to the case when F is a neighborly family, as follows: V(G(F)) = {1, ..., 15}, and i and j are connected by *one* edge for *every* value of k, such that $b_{i,j} \cdot b_{j,k} = -1$. Thus, G(F) is a multigraph, containing K_{15}. Let us define the graph H by the equation: $G(F) = K_{15} \oplus H$. The number of edges of H, denoted by m, is given by: $m = \sum_{i,j} ijx_{i,j} - 105$.

Case #	$x_{2,2}$	$x_{2,3}$	$x_{3,3}$	$x_{3,4}$	$x_{4,4}$	m
1.	1	2	1	0	5	0
2.	2	0	2	0	5	1
3.	2	1	0	1	5	1
4.	3	0	0	0	6	3
5.	1	0	0	0	7	11
6.	0	0	0	4	4	7
7.	0	0	1	2	5	8
8.	0	0	2	0	6	9
9.	0	1	0	1	6	9

Table 3. The nine solutions of the equations (7**, 8**).

The analogue of property (9) holds for G(F) in this case, too, namely:

$$G(F) = \sum_{i,j} x_{i,j} K_{i,j} .$$

In case 1, m = 0, hence $G(F) = K_{15}$. Its decomposition into *nine* complete bipartite graphs, namely $G(F) = K_{2,2}+ 2K_{2,3}+ K_{3,3}+5K_{4,4}$ is in contradiction with the Graham-Pollak Theorem. Therefore case 1 is impossible.

In cases 2 and 3, m = 1. Thus, G(F) is just a K_{15} plus an extra edge. The decomposition of G(F) into *nine* complete bipartite graphs contradicts Theorem 2 of [23]. This theorem states (for m = 1) that the graph K_n plus an edge as a multiple edge has no decompositions into fewer than n - 1 complete bipartite graphs. Thus cases 2 and 3 are impossible.

In case 4, $G(F) = 3K_{2,2} + 6K_{4,4} = K_{15}$ plus three edges. The tetrahedra of F are of types (4, 4, 4, 4) and (4, 4, 4, 2), hence the three extra edges form a triangle, say, (13, 14, 15).

Using earlier notations, let G(F) have the decomposition of the form:

$$G(F) = \sum_{i=1}^{9} K_{A_i,B_i} ,$$

and consider the following system:

$$x_{15} = 0$$

$$\sum_{i \in A_j} x_i = 0 , \quad 1 \leq j \leq 9$$

and

$$\sum_{i=1}^{15} x_i = 0 .$$

It follows, as in the proof of lemmas 14, 15 and 16, that $x_i = 0$ for all i; this is a contradiction, because there are fifteen variables and only eleven equations. Therefore case 4 is also impossible.

In the remaining cases, we analyse all the possibilities for H and show that Lemmas 13 - 16 lead to contradictions.

The maximum degree of H is at most two, and every connected component of H is either a path S_n on n edges, $n \geq 1$, or a cycle C_p, on p edges, where $p \geq 2$. Notice that digons are allowed and possible in H. Call a connected component *even* (*odd*) if it has an even (odd, respectively) number of edges. Let H_e (H_o) denote the subgraph of H, consisting of *all* the even (odd, respectively) connected components of H. Let |H| denote the number of *edges* of H, and let c(H) denote the number of connected components of H.

We have the following.

Lemmata: If $|H_e| = 2k$, then there exists a set W consisting of k vertices in H_e such that

$H_e \backslash W$ is empty; if $|H_o| = k$ and $c(H_o) = t$, then there exists a set W consisting of $(k-t)/2$ vertices, such that $H_o \backslash W$ consists of t independent edges.

Proof: In case $|H_e| = 2k$, let the set W be the set consisting of every other vertex on each cycle of H_e and every other vertex on the even paths of H_e , avoiding the end vertices. It follows that $H_e \backslash W$ is empty.

In the case of H_o, the set W is chosen in a similar way: in every cycle or path of odd length, say $2r + 1$, choose r vertices evenly dispersed and containing no end points on paths. Deleting these m vertices from that component leaves just one edge, hence $H_o \backslash W$ consists of t independent edges.

This completes the proof of the Lemmata.

In case 5, the only types of tetrahedra are $(4, 4, 4, 4)$ and $(4, 4, 4, 2)$, as only $x_{2,2}$ and $x_{4,4}$ do not vanish. Therefore all the vertices of H, where $G(F) = K_{15} \oplus H$, are two valent. As $|H| = 11$, $c(H_o)$ is odd. There are the following possibilities:

1. $c(H_o) = 1$; it follows by the Lemmata that for all the possibilities of $H = H_e + H_o$, there exists a set W of five vertices, such that $H \backslash W$ is an edge; the impossibility of this case is by Lemma 13.

2. $c(H_o) = 3$; it follows that either $H = C_5 +: C_3 +: C_3$ or else $H = C_3 +: C_3 +: C_3 +: C_2$; the impossibility is by Lemma 16, (i) and (ii).

Thus case 5 is impossible.

In case 6, $|H| = 7$ and $c(H_o)$ is odd, say $c(H_o) = 2q + 1$, $0 \le q \le 3$. Denote $|H_o| = 2r + 1$. Thus $|H_e| = 7 - (2r + 1) = 6 - 2r$. By the Lemmata there is a set W, consisting of $(2r + 1) - (2q + 1)]/2 = r - q$ vertices in H_o plus $3 - r$ vertices in H_e, such that $H \backslash W$ consists of q independent edges; W has $(r - q) + (3 - r) = 3 - q$ vertices. If $0 \le q \le 2$, then W can be enlarged to a set W_1 consisting of five vertices, such that $H \backslash W_1$ has one edge, or none, and Lemma 13 is applicable. If $q = 3$, then H consists of seven independent edges and Lemma 15 is applicable.

Therefore case 6 is impossible.

In case 7, $|H| = 8$, hence $c(H_o) = 2q$, $0 \le q \le 3$. Observe that $q \ge 4$ is impossible since H can have at most fifteen vertices. Let $|H_o| = 2r$. Thus by the Lemmata there is a set W, consisting of $(2r - 2q)/2 = r - q$ vertices from H_o and $(8 - 2r)/2 = 4 - r$ vertices from H_e; W has $4 - q$ vertices and $H \backslash W$ consists of 2q independent edges. If $0 \le q \le 2$, then W can be enlarged to a set W_1 of five vertices, such that $H \backslash W_1$ has one edge or none, hence Lemma 13 is applicable. If $q = 3$, then Lemma 15 is applicable (W has one vertex and $H \backslash W$ consists of six independent edges). Therefore case 7 is impossible.

In cases 8 and 9, $|H| = 9$, so that $c(H_o)$ is odd, say $c(H_o) = 2q + 1$, $0 \le q \le 3$ ($q \le 3$, since H has at most fifteen vertices). Let $|H_o| = 2r + 1$; by the Lemmata, there is a set W of $r - q$ vertices from H_o and $[9 - (2r + 1)]/2 = 4 - r$ vertices from H_e; W has $4 - q$ vertices and $H \backslash W$ consists of $2q + 1$ independent edges. If $0 \le q \le 1$, then W can be enlarged to a set W_1, consisting of five vertices, such that $H \backslash W_1$ has just one edge, or none, hence Lemma 13 is applicable. If $q = 2$, then W has two vertices and $H \backslash W$ consists of five independent edges, hence Lemma 14 is applicable.

If $q = 3$, $c(H_o) = 7$ and since H has at most fifteen vertices, H must be equal to $C_3 +: 6S_1$, implying that H has fifteen vertices. This cannot happen in case 9, since in this case $x_{2,3} = 1$, implying that three of the tetrahedra are of type (4, 4, 4, 2) and hence the corresponding three vertices of $G(H)$ have valence fourteen. Thus H has at most twelve vertices. Therefore $H = C_3 +: 6S_1$ and (being in case 8), $K_{15} \oplus H = 2K_{3,3} + 6K_{4,4}$; this is impossible by Lemma 16 (iii).

Thus cases 8 and 9 are impossible.

This completes the proof of Theorem 7.

Theorem 7 was used by Simon [14] in showing that $g(2, 4) \leq 14$, i.e., in showing that there can be at most *fourteen* nearly-neighborly quadrangles in the plane.

References

[1] Bagemihl, F., A conjecture concerning neighboring tetrahedra, Amer. Math. Monthly **63** (1956), 328-329.

[2] Baston, V.J.D., *Some properties of polyhedra in Euclidean Space*, Pergamon Press, Oxford, 1965.

[3] Besicovitch, A.S., On Crum's problem, J. London Math. Soc. **22** (1947), 285-287.

[4] Buck, R.C., Partitions of space, Amer. Math. Monthly **50** (1943), 541-544.

[5] Danzer, L., Grunbaum, B. and Klee, V. Jr., Helly's theorem and its relatives, Amer. Math. Soc. Proc. Symp. Pure Math. **7** (1963), 101-180.

[6] Furino, S., Gamble, B. and Zaks, J., Nearly-neighborly families of tetrahedra and the decomposition of some multigraphs, Res. Report. CORR 88-22, 1988, University of Waterloo, Canada.

[7] Graham, R.L. and Pollak, H.O., On the addressing problem for loop switching, Bell Sys. Tech. J. **50** (1971), 2495-2519; see also Springer Lecture Notes in Math., **303**, Springer, New York (1973), 99-110.

[8] Grunbaum, B., *Convex Polytopes*, J. Wiley & Sons, New York, 1967.

[9] Kassem, J., Neighborly families of boxes, Ph.D. thesis, The Hebrew University, Israel, 1985.

[10] Klee, V. Jr., Can nine tetrahedra form a neighboring family?, Amer. Math. Monthly **76** (1969), 178-179.

[11] Peck, G.M., New proof of a theorem of Graham and Pollak, Discrete Math. **49** (1984), 327-328.

[12] Perles, M.A., At most 2^{d+1} neighborly simplices in E^d, Proc. Conf. "Convexity and Graph Theory", Israel, 1981, Ann. Discrete Math. **20** (1984), 253-254.

[13] Reznick, B., Tiwari, P. and West, D.B., Decomposition of product of graphs into complete bipartite subgraphs, Discrete Math.**57** (1985), 189-193.

[14] Simon, J., Bounds on the cardinalities of nearly neighborly and neighborly families of polytopes, Ph.D. thesis, University of Illinois, Urbana, U.S.A., 1989.

[15] Tietze, H., Uber das problem der nachbargebiete im raum, Monatsh. Math. **16** (1905), 211-216.

[16] Tverberg, H., On the decomposition of K_n into complete bipartite graphs, J. Graph Theory **6** (1982), 493-494.

[17] Zaks, J., Bounds of neighborly families of convex polytopes, Geometriae Dedicata **8** (1979), 279-296.

[18] Zaks, J., Neighborly families of 2^d d-simplices in E^d, Geometriae Dedicata **11** (1981), 505-507.

[19] Zaks, J., Arbitrarily large neighborly families of symmetric convex polytopes, Geometriae Dedicata **20** (1986), 175-179.

[20] Zaks, J., How does a complete graph split into bipartite graphs and how are neighborly cubes arranged, Amer. Math. Monthly **92** (1985), 568-571.

[21] Zaks, J., A solution to Bagemihl's conjecture, C. R. Math. Rep. Acad. Sci. Canada **8** (1986), 317-321.

[22] Zaks, J., Neighborly families of congruent convex polytopes, Amer. Math. Monthly **94** (1987), 151-155.

[23] Zaks, J., Nearly-neighborly families of tetrahedra and the decomposition of some multigraphs, J. Comb. Theory A. **48** (1988), 147-155.

University of Haifa, Israel.

MEMOIRS of the American Mathematical Society

SUBMISSION. This journal is designed particularly for long research papers (and groups of cognate papers) in pure and applied mathematics. The papers, in general, are longer than those in the TRANSACTIONS of the American Mathematical Society, with which it shares an editorial committee. Mathematical papers intended for publication in the Memoirs should be addressed to one of the editors:

General instructions to authors for

PREPARING REPRODUCTION COPY FOR MEMOIRS

For more detailed instructions send for AMS booklet, "A Guide for Authors of Memoirs." Write to Editorial Offices, American Mathematical Society, P.O. Box 6248, Providence, R.I. 02940.

MEMOIRS are printed by photo-offset from camera copy fully prepared by the author. This means that the finished book will look exactly like the copy submitted. Thus the author will want to use a good quality typewriter with a new, medium-inked black ribbon, and submit clean copy on the appropriate model paper.

Model Paper, provided at no cost by the AMS, is paper marked with blue lines that confine the copy to the appropriate size.

Special Characters may be filled in carefully freehand, using dense black ink, or **INSTANT** ("rub-on") **LETTERING** may be used. These may be available at a local art supply store.

Diagrams may be drawn in black ink either directly on the model sheet, or on a separate sheet and pasted with rubber cement into spaces left for them in the text. Ballpoint pen is not acceptable.

Page Headings (Running Heads) should be centered, in CAPITAL LETTERS (preferably), at the top of the page — just above the blue line and touching it.

LEFT-hand, EVEN-numbered pages should be headed with the AUTHOR'S NAME;

RIGHT-hand, ODD-numbered pages should be headed with the TITLE of the paper (in shortened form if necessary).

Exceptions: PAGE 1 and any other page that carries a display title require NO RUNNING HEADS.

Page Numbers should be at the top of the page, on the same line with the running heads.

LEFT-hand, EVEN numbers — flush with left margin;

RIGHT-hand, ODD numbers — flush with right margin.

Exceptions: PAGE 1 and any other page that carries a display title should have page number, centered below the text, on blue line provided.

FRONT MATTER PAGES should be numbered with Roman numerals (lower case), positioned below text in same manner as described above.

MEMOIRS FORMAT

It is suggested that the material be arranged in pages as indicated below. Note: Starred items (*) are requirements of publication.

Front Matter (first pages in book, preceding main body of text).

Page i — *Title, *Author's name.

Page iii — Table of contents.

Page iv — *Abstract (at least 1 sentence and at most 300 words).

Key words and phrases, if desired. (A list which covers the content of the paper adequately enough to be useful for an information retrieval system.)

**1991 Mathematics Subject Classification.* This classification represents the primary and secondary subjects of the paper, and the scheme can be found in Annual Subject Indexes of MATHEMATICAL REVIEWS beginnning in 1990.

Page 1 — Preface, introduction, or any other matter not belonging in body of text.

Footnotes: *Received by the editor date.
Support information — grants, credits, etc.

First Page Following Introduction – Chapter Title (dropped 1 inch from top line, and centered). Beginning of Text.

Last Page (at bottom) – Author's affiliation.